36th Eg-MRS International Conference

IOP Conference Series: Materials Science and Engineering
Volume 1269

Cairo, Egypt
24 – 25 September 2022

ISBN: 978-1-7138-6428-8
ISSN: 1757-8981

Printed from e-media with permission by:

Curran Associates, Inc.
57 Morehouse Lane
Red Hook, NY 12571

proceedings.com

Some format issues inherent in the e-media version may also appear in this print version.

Institute of Physics
Dirac House, Temple Back
Bristol BS1 6BE UK

Phone: 44 1 17 929 7481
Fax: 44 1 17 920 0979

techtracking@iop.org

Additional copies of this publication are available from:

Curran Associates, Inc.
57 Morehouse Lane
Red Hook, NY 12571 USA
Phone: 845-758-0400
Fax: 845-758-2633
Email: curran@proceedings.com
Web: www.proceedings.com

TABLE OF CONTENTS

IOP Conf. Series: Materials Science and Engineering 1269 (2022) 011001 doi:10.1088/1757-899X/1269/1/011001

Proceeding of 36th International Conference of Material Sciences and Its Applications (36th Eg-MRS) 2022

Preface

Arafa H.Aly
Physics department, Faculty of Sciences, Beni Suef University

It is with great pleasure that I introduce the proceedings of the 36th Eg-MRS International Conference, 24-25 September 2022, Cairo, Egypt. This conference was dedicated to technical issue related to material sciences research and its applications.

The objectives of the conference were not only to provide opportunities to the international scientists to present state of the art research in material sciences but also to provide a forum to discuss ideas for future directions in the field and explore possibilities for collaborative research projects.

The conference program included keynote, oral, and poster presentations from scholars working in the areas of materials science and engineering from all over the world. It covered recent trends and progress made in the field of material sciences. Professors from Egypt, Malaysia, USA, Canada, India, Russia and France were invited to deliver keynote lectures regarding the latest information in their respective areas of expertise.

All papers were subjected to peer-review by conference committee members and international reviewers. The papers were selected based on their quality and their relevance to the conference. The proceedings present recent advances in the fields of materials science and engineering and manufacturing as well as in other related areas including materials properties, measuring methods, analytical and numerical modeling and simulations among others. I would like to express special gratitude to members of the conference committee and organizers of the conference. I would also like to thank the reviewers for their valuable time and advice which helped in improving the quality of the papers selected for presentation at the conference and for publication in the proceedings. Finally I want to thank the authors, the members of the organizing committee, the reviewers, the chairpersons, sponsors, and all other conference participants for their support of Eg-MRS 2022.

IOP Conf. Series: Materials Science and Engineering 1269 (2022) 011001 doi:10.1088/1757-899X/1269/1/011001

Conference Chairman

Prof. Dr Arafa Hussien Aly
Beni-Suef University, Egypt
Eg-MRS Chiarman

Coordinator (Eg-MRS)

Prof Dr.Hassan Talaat (Ain Shams University, Egypt)
Prof. Dr. Hassan Afify *(National Research Center, Egypt)*
Prof. Dr.Talaat Meaz *(Tanta University, Egypt)*
Dr. Hassan Nageh *(The British University in Egypt)*
Prof Tamer Abdalla (Ain Shams University, Egypt)

Organizing and Scientific Committee

Prof Dr. Ahmed Ramadan *(Helwan University*
Prof Dr. Sohair Negm *(Benha University)*
Prof Dr. Hamed Hassan *(Alexandria University)*
Dr. Hassan Nageh (BUE)
Prof Dr. Samia A Saafan *(Tanta University)*
Prof Dr. Mohamed Shaban (Beni-Suef Univ.)
Prof Dr. Ahmed Ghitas (NRIAG)
Prof Dr. Soltan Eid A. Soltan *(Helwan Uiversity)*
Dr. Mohmed Saleh (MUST)
Mr. Zaky A Zaky (Beni Suef University)

Editors and Reviewers

Editor Prof Dr.Arafa Hussien Aly, Beni-Suef Univerrsity, Egypt

Reviewers

1. Prof Dr.Hassan Talaat, Ain Shams University, Egypt
2. Prof Dr. Hassan Afify, National Research Center, Egypt
3. Prof Dr.Adel Phillips, Ain Shams University, Egypt
4. Prof Dr.Hamdy Farghaly, El-Minia University, Egypt
5. Prof Dr.Sedky Sodky Abolila , Sohag University, Egypt
6. Prof Dr. Iman Mahdy, Al-Azhar University, Egypt
7. Prof Arafa H. Aly, Beni Suef University, egypt
8. Prof. Tarek El-brolossy, Ain Shams University, Egypt
9. Dr.Tallat Abdel Hamid, National Research Center, Egypt
10. Prof Walid Tawfik , Cairo University, Egypt
11. Prof Mahi Singh, Canada
12. Prof Ahmed El-Badawi, Minia University, E
13. Prof Saeed AbdelGhany, Benha University, Egypt
14. Prof Sahar Elnagar, Cairo University, Egypt
15. Dr.Hassan Nageh, BUE, Egypt

IOP Conf. Series: Materials Science and Engineering 1269 (2022) 011002 doi:10.1088/1757-899X/1269/1/011002

Overview of the 36th International Conference, 36th Eg-MRS 202

Arafa H.Aly

Physics Department, Faculty of Sciences, Beni-Suef University, Egypt

arafaaly@aucegypt.edu

Abstract. The overview describes the main directions and results of the 36th Eg-MRS International Conference, 36th Eg-MRS-2022 held in Cairo, Egypt, 24-25 September 2022. It gives the details about the participants and the proceedings.

The Volume contains Proceedings of the 36th Eg-MRS International Conference, Eg-MRS-2022:

Modernization, Innovations and Progress" which was held in Cairo, Egypt in September 24-25, 2022.. The Conference held online using Zoom due to COVID-19.

The purpose of the Conference is to share the results and prospects of the achievements in using advanced scientific, innovative and information technologies in materials sciences and its applications.

In fact, this is the third year of publishing a special issue of Eg-MRS international conferences under the umbrella of IOP material sciences and engineering. The first one was in November 2019; Hurghada, Egypt [1]. The second and third was in August 2020 & July 2022 which was online due to COVID-19 [2,3]

All conference organizers/editors are required to declare details about their peer review. Therefore, please provide the following information:

- **Type of peer review: Single-blind / Double-blind / Triple-blind / Open / Other (please describe) Double-blind**
- **Conference submission management system: Online: https://cutt.us/mo8G5**
- **Number of submissions received: 50**
- **Number of submissions sent for review: 30**
- **Number of submissions accepted: 11**
- **Acceptance Rate (Number of Submissions Accepted / Number of Submissions Received X 100): 22%**
- **Average number of reviews per paper: 3**
- **Total number of reviewers involved: 20**
- **Any additional info on review process: No**

- **Contact person for queries:** Prof Arafa H.Aly; **arafaaly@aucegypt.edu**

Please submit this form along with the rest of your files on the submission date written in your publishing agreement.

The information you provide will be published as part of your proceedings.

[1] Arafa H. Aly, 2020 *IOP Conf. Ser.: Mater. Sci. Eng.* **762, 0122005**

[2] Arafa H. Aly, 2020 *IOP Conf. Ser.: Mater. Sci. Eng.* **956, 011002**

[3] Arafa H.Aly, 2021 *IOP Conf. Ser.: Mater. Sci. Eng.* **1171**, **011001**

SnO_2 Doping Effect on the Microstructural and Electrical Behavior of ZnO nanoparticles-Bi_2O_3 Based Varistor Ceramics

R. K. Sendi
Umm Al- Qura University (UQU), Faculty of Applied Science, Physics Department
21955 Makkah Al- Mukarramah, Saudi Arabia
E-mail: rksendi@uqu.edu.sa

Abstract. In the current study, ZnO-Bi_2O_3 varistors manufactured from ZnO nanoparticle size and doped with SnO_2 were processed by the ceramic industrial technique. The impact of the various SnO_2 concentration upon the sintering, microstructural improvement and nonlinearity characteristic of the varistor system was investigated. Different SnO_2 concentration was found to possess considerable influences on the varistors, particularly on improving ZnO grains growth at a lower SnO_2 amount. But, the adding of greater contents of SnO_2 leads to inhibit the ZnO grain growth via the raising percentage of Zn_2SnO_4 type spinel phase. In discs with a Sn/Bi less than 1.5 mol%, the pyrochlore phase forms and bounds all the SnO_2, whereas excess Bi_2O_3 leads to the formation of a liquid phase at 740 °C which promotes sintering process. For Sn/Bi more than 1.5 mol%, all the Bi_2O_3 is bounded in $Bi_2Sn_2O_7$ pyrochlore phase and sintering process is prevented. X-ray diffraction investigated that additive of SnO_2 to the ZnO-Bi_2O_3 varistor sample caused the structure of the $Bi_2Sn_2O_7$ type pyrochlore phase and the Zn_2SnO_4 type spinel phase during the sintering step. The SnO_2 adding furthermore noticeably affected the electrical behaviors of the sample with a significant rise in the breakdown voltage with increasing the SnO_2 doping more than 1.5 mol%. The resistivity also achieved an obvious increase with increase the SnO_2 doping contents. Moderate SnO_2 doping enhanced the nonlinear properties of the ZnO-Bi_2O_3 varistor system, while a significant amount of SnO_2 causing its deterioration.

1. Introduction

Many studies have investigated the impacts of processing and microstructure characteristics on various electronic and electrical behaviors varistor system [1, 2]. It has been investigated that ZnO varistors with heterogeneous structures continually experience a critical perversion of the I-V behaviors owning to huge overloads current attributed to single massive ZnO grains, which caused immediate the ZnO varistor degradation in electrical and electronic processes.

The addition of Bi_2O_3 to varistor is fundamental for the non-ohmic behaviour to occur (The nonlinear coefficient α below 10). In the sintering process, Bi_2O_3 provides a liquid phase into the varistor system and creates a Bi-rich phases around the ZnO grains. The metal oxides addition such as Cr_2O_3, Sb_2O_3, Mn_2O_3, Co_3O_4 and others enhances the electrical behaviors of the basic binary ZnO-Bi_2O_3 varistor system.

The influence of SnO_2 on ZnO-based varistor made from micro ZnO powder has been reported by Pillai, S. C. et al. [3]. They investigated that the addition of SnO_2 increases the nonlinear exponent (α up to 20), electrical resistivity in the prebreak down region, and enhances linear shrinkage of the basic ZnO-Bi_2O_3 binary varistor system.

2. Experimental

2.1 Sample preparation

The raw materials applied in this work were of significant purity and quality. Below are the raw materials list and their corresponding properties. The technical specifications of the raw materials are posted in following Table 1.

Material	Chemical formula	Purity (%)	Mean size	Supplier
Nanoparticles Zinc Oxide	ZnO	> 99.9	20 nm	MKNANO Canada
Bismuth Trioxide	Bi_2O_3	> 99.9	10 µm	SEGMA-ALDAICH
Tin (IV) Oxide	SnO_2	> 99.9	0.7 µm	SEGMA-ALDAICH

Table 1: The sizes of ZnO, Bi_2O_3, SnO_2, grains

ZnO-Bi_2O_3 varistors with the following constitution were prepared:
(99.5-x) mol% 20-nm ZnO + 0.5 mol% Bi_2O_3 + x mol% SnO_2 (x = 0.0, 0.5, 1.0, 1.5, 2.0, 2.5, 3.0).
For each disc, the powder was blended with polyvinyl alcohol (PVA). The resulting granules were applied to form discs by pressing at 4 ton/cm^2 pressure. Finally, the green discs were sintered at 1200 oC for 3 h. To obtain the electrical characteristics, silver pastes were coated and toasted on both sides of the sintered discs.

2.2 Characterization
The apparent and relative density of the sample was measured by dimensional and Archimedean methods, respectively. For the Archimedes method, a Mettler balance (Type AE200) with a particular density measurement device for solid materials (Type AB33360) was utilized. The measuring principle demonstrates the evaluation of the mass in normal air and in the fluid. The balance has a measuring accuracy of ±0.1 mg. The density is obtained from the following equation:

$$\rho = \frac{m_a}{m_a - m_{fl}} \cdot \rho_{fl}$$

ρ_{fl} is the density of the fluid, which is temperature dependent, m_a is the sample mass in air and m_{fl} is the sample mass in the fluid.
The microstructure of the ZnO varistors was investigated via scanning electron microscopy (SEM) (model: JSM – 6460 LV). The average grain size (D) was estimated with the linear intercept technique [4-6].
The crystalline phases were considered by a significant resolution X-ray diffractometer (XRD) system (PANalytical X' Pert PRO MED PW6023) with λ=1.5406Å Kα radiation.

2.3 Electrical testing
The I – V properties of the discs were measured using a significant voltage source (KEITHLEY instruments 246 high voltage supply). The capacitance–voltage (C–V) properties of the varistors were determined at 1 kHz by applying Keithley 6410 recorder.

3. Result and Discussion
3.1 Physical analysis
The green densities of all samples with different amounts of SnO_2 are 4.11 ± 0.1 g/cm^3. However, the sintered density of the samples was in the range of 91%–95% of the theoretical density. A sintered density of more than 95% of the theoretical density is difficult to obtain because zinc nanoparticles tend to vaporize in the sintering process.
Figure 1 exhibits the differences of the apparent and relative densities of the varistor ceramics with various SnO_2 doping amounts. The apparent density of the varistors also raised obviously with the SnO_2 adding amounts up to 1.5 mol%, which may be attributed to the greater density of SnO_2 (6.995 g/cm^3) than that of ZnO (5.672 g/cm^3) as reported by Feng H., et al. [7], and a few raise of relative density also observed (Figure 1). The huge ZnO discs densification at the high SnO_2 doping contents might be attributed to the SnO_2 doping impacts within the grain lattice, which led to the oxygen vacancies formation during the sintering step.

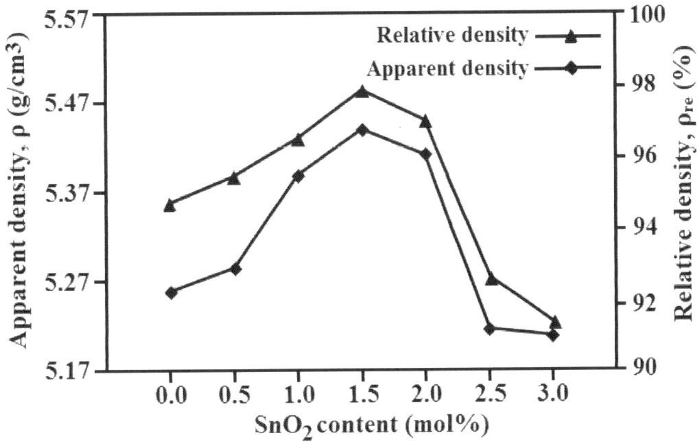

Figure 1. The apparent and relative densities of ZnO nanoparticles–Bi_2O_3 based varistor ceramics doped with various amounts of SnO_2 additive

3.2 Microstructural analysis

3.2.1 Scanning electron microscopy

Structural characteristics of ZnO nanoparticle-Bi_2O_3 based varistors doped with various contents of SnO_2 amounts are obtained via SEM analysis (Figure 2) and the sizes of ZnO grain are calculated in Table 2. These SEM micrographs indicate that the morphology of disc surface was significantly affected by the SnO_2 adding compositions, and the mean size of grain changed with various doping amounts.

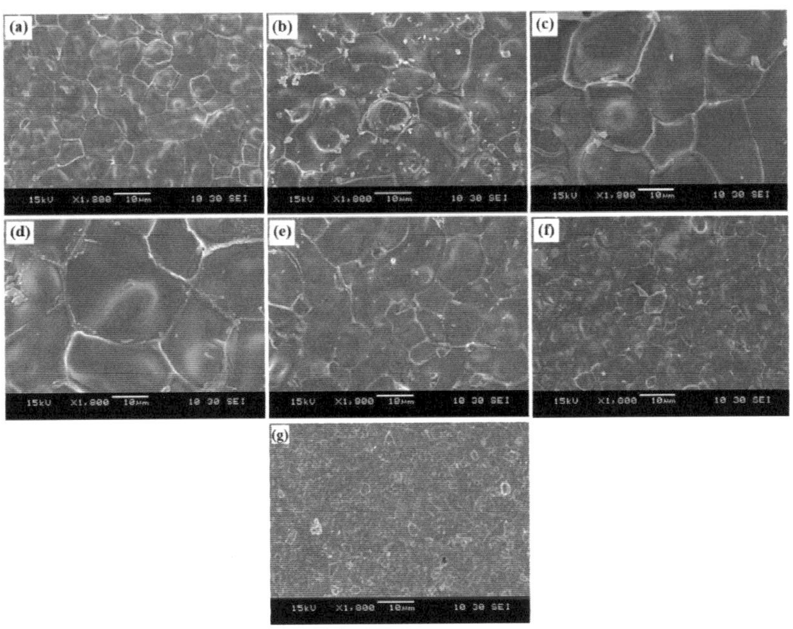

Figure 2. Typical SEM micrographs of ZnO nanoparticles–Bi2O3 based varistor ceramics doped with various amounts of SnO_2 additive: (a) 0.0 mol%, (b) 0.5 mol%, (c) 1.0 mol%, (d) 1.5 mol%, (e) 2.0

mol%, 2.5 mol% and 3.0 mol%

An important notice is the obvious presence of considerable amount of $Z_{n2}SnO_4$ type spinel phase (1–4 μm) in sample with elevated SnO_2 doping amount. While in varistor with less doping contents, the Bi_2O_3 phases form a thin layer at the ZnO grain boundaries. The SnO_2 added to the ZnO nanoparticles-Bi_2O_3 system results in pyrochlore and spinel phases formation. The Zn_2SnO_4 type spinel phase form in the sample according to the reaction [8]:

$$2ZnO + SnO_2 \xrightarrow{Sintering\ process} Zn_2SnO_4 \quad (1)$$

The $Bi_2Sn_2O_7$ type pyrochlore phase [8] formed in the Bi_2O_3-SnO_2 as well as in the presence of ZnO, according to the following equation:

$$Bi_2O_3 + 2SnO_2 \xrightarrow{High\ temperature} Bi_2Sn_2O_7 \quad (2)$$

In the presence of ZnO, the pyrochlore phase ($Bi_2Sn_2O_7$), which forms during cooling in the ZnO-Bi_2O_3-SnO_2 varistor, decomposes to the Zn_2SnO_4 spinel phase and Bi_2O_3 liquid phase, according to the following equation [8]

$$Bi_2Sn_2O_7 + 4ZnO \xrightarrow{High\ temperature} 2Zn_2SnO_4 + Bi_2O_3 \quad (3)$$

The existence of secondary phases within the samples possesses a significant effect on the growth of the grain. In the varistor without any SnO_2 added, the development of grain is enhanced and the porosity predominates with the existence of the Bi_2O_3-liquid phase. The addendum of a little content of SnO_2 (less than 1.5 mol%) in the varistor leads to a significantly huge size of grains with minimal intragranular porosity compared with the previous disc (0.0 mol% of SnO_2 doping). The addition of a lower amount of SnO_2 (less than 1.5 mol%) only leads to a small amount of formation of Zn_2SnO_4 type spinel phase which does not impact significantly the growth of ZnO grains. Therefore, increased growth of ZnO grain within the varistors may be attributed to the impact of SnO_2 on the Bi_2O_3 liquid phase wettability. The size of the grain was reduced with the augmentation of doping contents (more than 1.5 mol%). The decrease in grain growth may be caused by the increasing Zn_2SnO_4 spinel phases at the boundaries through the sintering step [9,10].

An interesting observation is a clear presence of some spinel phases (1–4 μm), which may impede the growth of the ZnO grains during the sintering process. These spinel phases are usually located in groups but they may also be found individually between, sometimes even within, the ZnO grains. The liquid phase during sintering produces Bi-rich phases that are associated with the spinel particles in the intergranular region of the microstructure.

The clear presence of white spots in Figure 2 (b) may be attributed to bismuth at the grain boundary (GB), between the grains, where no bismuth is detected in the ZnO grains. Large numbers of Bi-ions with large ionic sizes may have been segregated to the GB. These Bi-ions could have encountered difficulties in occupying interstitial sites inside the ZnO grains because of the occupation of the extra donor, and they ended up at the GB to form Bi-rich GB phases because they could not squeeze into the lattice atoms because of their large sizes.

IOP Conf. Series: Materials Science and Engineering 1269 (2022) 012001 doi:10.1088/1757-899X/1269/1/012001

Table 2. The Microstructural and physical parameters of the ZnO nanoparticles–Bi_2O_3 based varistor ceramics doped with various amounts of SnO_2 additive

Samples	ρ (g/cm³)	d (μm)	Diameter shrinkage %
0.0	5.259	21.7	18.326
0.5	5.285	25.6	17.744
1.0	5.389	29.4	17.520
1.5	5.440	31.7	16.973
2.0	5.412	28.2	15.496
2.5	5.216	19.4	15.038
3.0	5.209	11.3	14.951

3.2.2 X-ray diffraction

Figure 3 explains the X-ray diffraction (XRD) spectra of the pure ZnO nanoparticle–Bi_2O_3 varistor doped with various amounts of SnO_2 distinguished by the highest main peaks that formed from the ZnO layer. Other peaks, namely, α-Bi_2O_3 phase, $Bi_2Sn_2O_7$ type pyrochlore phase and Zn_2SnO_4 type spinel phase appeared as secondary phases.

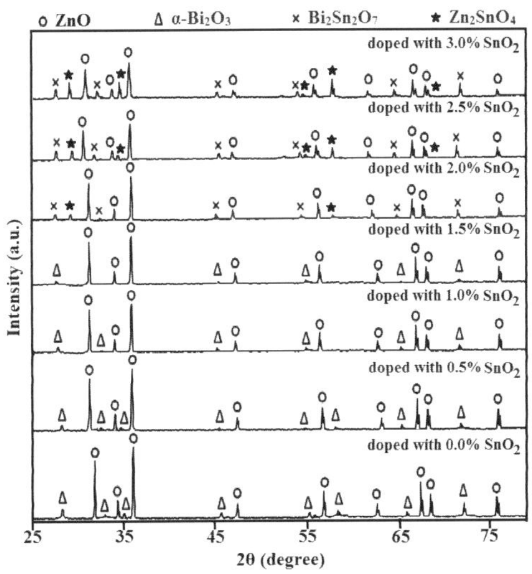

Figure 3. XRD patterns of ZnO nanoparticles–Bi_2O_3 based varistor ceramics doped with various amounts of SnO_2 additive

Figure 3, shows that the phase of the pure varistor is ZnO and α-Bi_2O_3 phases. With augmentation SnO_2 content up to 1.5 mol%, the intensities of ZnO and α-Bi_2O_3 peaks reduce gradually but there is no more phase observed. However, with the increasing SnO_2 amounts up to 3.0 mol%, a new phase, $Bi_2Sn_2O_7$ type pyrochlore phase, was detected, and α-Bi_2O_3 peak did not appear at XRD spectra. Consequently, it could be deduced that the SnO_2 will react with Bi_2O_3 through the sintering step, forming $Bi_2Sn_2O_7$ type pyrochlore phase. The presence of the Zn2SnO4 type spinel phase is evident from XRD patterns in all SnO_2 doped varistors, except in a sample with 0.5 mol% SnO_2 doping amount.

3.3 Electrical properties
3.3.1 C-V characteristics

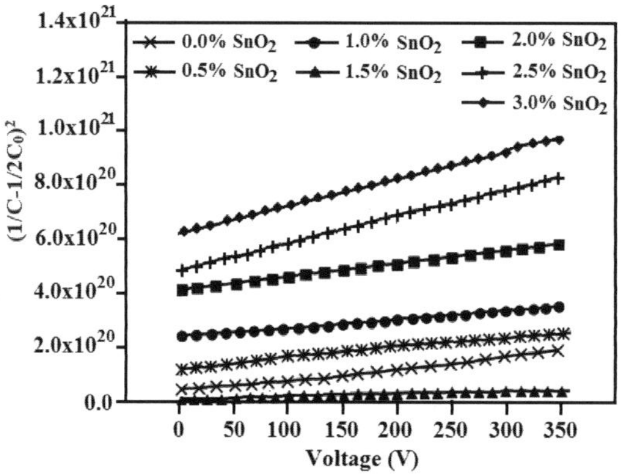

Figure 4. The C–V characteristics of ZnO nanoparticles–Bi_2O_3 based varistor ceramics doped with various amounts of SnO_2 additive

Figure 4 explains the C–V characteristics of varistors manufactured from 20nm ZnO powder and doped with various contents of SnO_2. The C–V behavior parameters of the varistors involving interfacial states density (N_{IS}), donor density (N_d), barrier height (Φ_B) and depletion layer (ω) were obtained, and the results are displayed in Table 2 and illustrated in Figure 5.

The density of donor (N_d) and the height of the barrier (Φ_B) were calculated using the following equation [11].

$$\left(\frac{1}{C}-\frac{1}{2C_o}\right)^2=\frac{2t}{A^2 de\varepsilon N_d}V+\frac{2t^2}{A^2d^2e\varepsilon N_d}\Phi_B \quad (4)$$

where C and C_o are the capacitances of the varistors at various voltages, t is the varistor thickness, A is the area of electrode, d is the average size of ZnO grain, ε is the permittivity of ZnO and e is charge of electron.

The barrier height $\emptyset_B$ may be determined from a plot of $1/C^2$ as a function of applied reverse bias and it is given by a following equation [12]:

$$\emptyset_B = V_{bi} + \varphi_n + \frac{kT}{q} - \Delta\Phi \quad (5)$$

Where $\Delta\Phi$ is the potential change due to the image force, and φ_n is given by the equation:

$$\varphi_n = \frac{kT}{q} lin \frac{N_D - N_A}{N_C} \quad (6)$$

Where N_D and N_A are the concentrations of donors and acceptors (in the n and p sides), respectively, and the effective density of states, N_c is given by:

$$N_C = 2\left(\frac{2\pi m^* kT}{h^2}\right)^{\frac{3}{2}} \quad (7)$$

Where m^* is the electron effective mass and h is Planck's constant.

The depletion layer (ω) is obtained from the following relationship [13]:

$$\omega = \left[\frac{2\epsilon_r\epsilon_0}{q}\left(\frac{N_A+N_D}{N_A N_D}\right)(V_{bi} - V)\right]^{\frac{1}{2}} \quad (8)$$

Where ϵ_r is the relative dielectric permittivity of the semiconductor, V_{bi} is the built-in voltage, and V is the applied bias.

N_d of the varistors doped with SnO_2 was promoted. Thus, it is supposed that the SnO_2 addition behaves like a donor within 20nm ZnO–Bi_2O_3 varistor ceramics [14, 15]. N_d increment is related to a reduction of interfacial states density N_{IS}, also depletion layer ω lowering. The varistors barrier height (Φ_B) enhances first and then drops with rising SnO_2 doping amounts, where the varistors doped with 0.5, 1.0 and 1.5 mol% represent proportional greater height of barrier compared to other samples with various SnO_2 doping contents. The values of barrier height Φ_B are calculated based on the fact that all the grains are considered electroactive. Due to the rise of barrier height in favor of the increase of varistor's nonlinear coefficient [15], so it can be predicted that the varistors doped with 0.5 mol%, 1.0 mol% and 1.5 mol% SnO_2 would have a relative higher nonlinear coefficient as shown in Table 3. For the varistors with SnO_2 doping contents more than 1.5 mol%, the increment in the capacitance with bias voltage can be associated with degradation in the nonohmic properties including a distortion of the Schottky barriers at intergrains. An alteration in the donor concentration would directly impact the capacity and conductivity of the varistor samples.

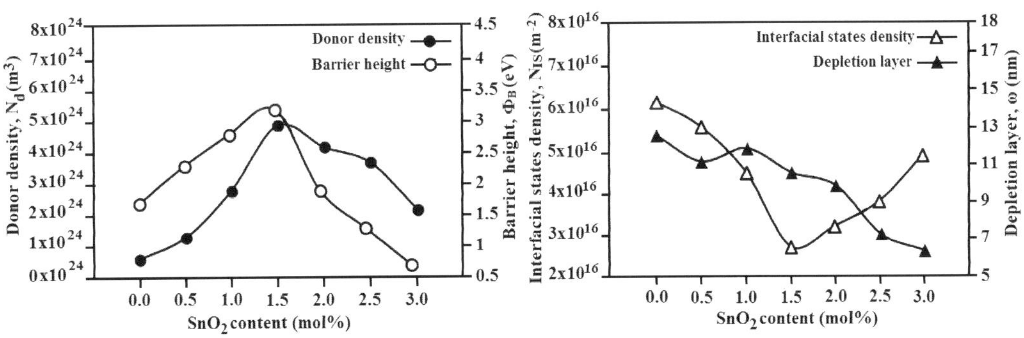

Figure 5. The C–V behavior parameters including: (a) donor densities and barrier heights, and (b) interfacial states density and depletion layer of ZnO nanoparticles–Bi_2O_3 based varistor doped with various amounts of SnO_2 additive

3.3.2 V-I behavior

Figure 6 displays the voltage/current behaviors of the 20nm ZnO–Bi_2O_3 varistors manufactured from 20nm ZnO and doped with various SnO_2 contents. The nonlinearity coefficient (α) can be investigated by using the following formula [16]

$$\alpha = \frac{\log\,(I_2 - I_1)}{\log\,(V_2 - V_1)} \quad (9)$$

Where α is nonlinear coefficient. I_1 and I_2 are the electric current corresponding to the electric voltage V_1 and V_2, respectively.

The different parameters of the electrical properties determined from the V-I graph are mentioned in Table 3. Doping with SnO_2 enhanced the non-linear property and the maximum value observed in the disc with addition SnO_2 is no more than 1.5 mol%. A significant SnO_2 amount leads to a degradation of the nonlinear behaviors. Therefore, SnO_2 has a considerable impact upon the height of barrier within the grain boundary.

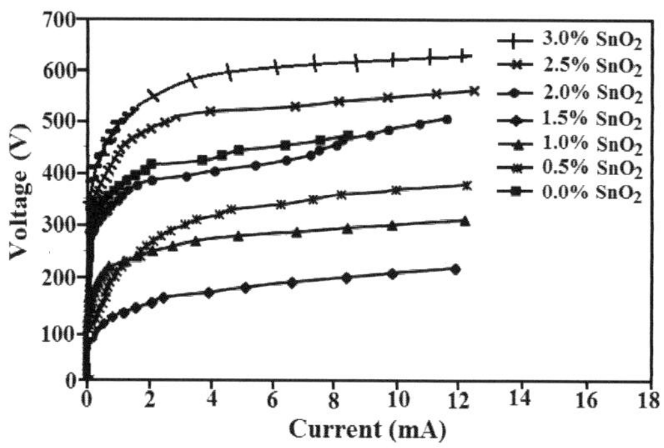

Figure 6. The Voltage-Current characteristic of ZnO nanoparticles–Bi_2O_3 based ceramic varistors doped with various amounts of SnO_2 additive

When the SnO_2 adding contents are exceeding 1.5 mol%, as presented in the XRD results in Figure 3, the $Bi_2Sn_2O_7$ phase amount increases obviously and α-Bi_2O_3 phase is absent. This may be attributed to the formation of $Bi_2Sn_2O_7$ will exhaust Bi_2O_3 during the sintering process.

With increasing SnO_2 doping amounts to more than 1.5 mol%, the sample breakdown voltage increments. These values of breakdown voltage are related to the grain size and the used electric voltage per boundary among the grains [17, 18], as presented in Figure 7. The rising in V_b could be explained via the reduction of the size of ZnO grains with a further rise in the SnO_2 adding content, and hence increasing the grain boundaries numbers among the electrodes which leads to raised "p-n junctions".

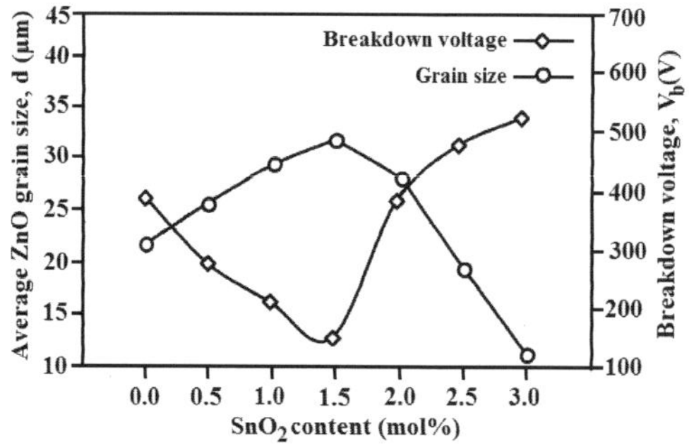

Figure 7. The variation of average ZnO grain sizes and breakdown voltage of ZnO nanoparticles–Bi_2O_3 based varistor with different amounts of SnO_2 additive

The resistivity at varistors with SnO_2 adding more than 1.5 mol% increased obviously. We proposed that the augmentation in the sample resistivity may be caused by increasing the barrier height at the boundaries attributed to a little growth of ZnO grains after increasing SnO_2 adding contents.

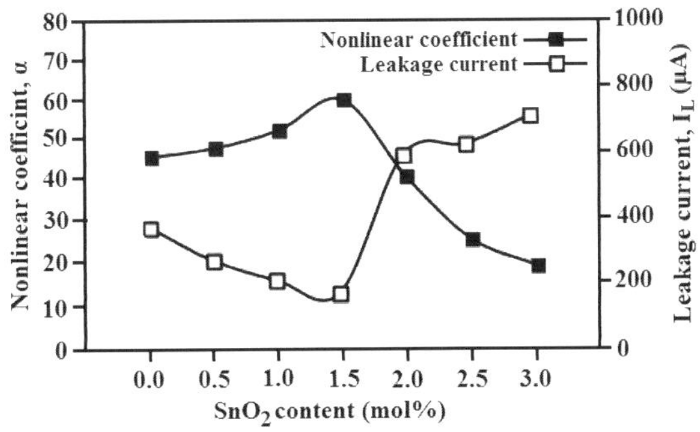

Figure 8. The variation of nonlinear coefficient and leakage current of ZnO nanoparticles–Bi_2O_3 based varistor with different amounts of SnO_2 additive

Figure 8 indicated that the reduce of the leakage currents caused by the rise of their nonlinearity coefficients; furthermore, at high SnO_2 addition content, the sudden development of the leakage currents perhaps caused by the sharp reduce of the varistors relative densities [18].

Table 3. The Summarizes the current-voltage (I-V) and capacitance-voltage (C-V) characteristic parameters of the ZnO nanoparticles–Bi_2O_3 based varistor ceramics doped with various amounts of SnO_2 additive

Doping amount of SnO_2 (mol%)	V_b (V)	α	ρ (kΩ.cm)	I_L (μA)	V_{gb} (V)	N_d (m^3) ($x10^{24}$)	N_{IS} (m^{-2}) ($x10^{16}$)	Φ_B (eV)	ω (nm)
0.0	390	44	302.5	374	5.44	0.6	6.2	1.7	12.6
0.5	276	46	286.9	274	5.63	1.3	5.6	2.3	11.2
1.0	210	50	226.3	214	5.78	2.8	4.5	2.8	11.9
1.5	149	57	139.4	175	5.96	4.9	2.7	3.2	10.6
2.0	385	39	312.8	597	3.75	4.2	3.2	1.9	9.9
2.5	481	25	493.3	629	1.29	3.7	3.8	1.3	7.3
3.0	527	19	627.1	714	0.89	2.2	4.9	0.7	6.4

4. Conclusion

ZnO-Bi_2O_3 varistors manufactured from ZnO nanoparticle size and doped with SnO_2 were processed. The impact of the various SnO_2 concentration on the sintering, microstructural and electrical behaviors of the varistor system was investigated. The mean size of grains is reduced accompanied with increasing SnO_2 contents causing a fundamental increase in the value of breakdown voltage. The SnO_2 acted as donor within varistor system, and SnO_2 content up to 1.5 mol% may enhance the nonlinearity coefficient of the samples. Therefore, integration of SnO_2 is likely to be a way to obtain a rising electric field ZnO varistor substances for possibility cost-efficient usage.

References

[1] Liu, M., Demeritte, T., Shankar, V., and Dai, Q. 2017 *Nanosci. Nanotechnol. Lett.* **9** (3) 388-391.

[2] Hembram, K., Rao, T. N., Srinivasa, R. S., and Kulkarni, A. R. 2015 *J. Eur. Ceram. Soc.* **35**

(13) 3535-3544.
[3] Pillai, S. C., Kelly, J. M., Ramesh, R., and McCormack, D. E. 2013 *J. Mater. Chem.***1** (20) 3268-3281.
[4] Sousa, V. C., Oliveira, M. M., Orlandi, M. O., and Longo, E. 2010 *J. Mater. Sci.:Mater. Electron.* **21** (3) 246-251.
[5] Kharchouche, F. 2018 *J. Mater. Sci.:Mater. Electron.* **29** (5) 3891-3897.
[6] Bouchekhlal, A., and Hobar, F. 2018 *J. Adv. Dielectr.* **8** (02) 1850014.
[7] Feng, H., et al. 2011 *J. Alloy. Comp.* **509** 7175-7180.
[8] Tomnic, S., Recnik, A., Bernik, S., Mazaj, M., Spreitzer, M., and Daneu, N. 2020 *J. Eur. Ceram. Soc.* **40** (15) 5518-5522.
[9] Nahm, C.W. 2010 *Mater. Sci. Eng. B.* **170** (1) 123-128.
[10] Nahm, C.W. 2010 *Ceram. Inter.* **36** (5) 1495-1501.
[11] Fu, Z., He, J., Lu, J., Fang, Z., and Wang, B. 2019 *Ceram. Inter.* **45** (17) 21900-21909.
[12] Lyle, L. A., Jiang, L., Das, K. K., and Porter, L. M. 2019 *Gallium Oxide* 231-261.
[13] Sasikala, B., Khan, A., Pooranchandra, S., Sasikala, B. 2005 Introduction to Electrical , Electronics and Communication Engineering.
[14] Roy, S., Roy, T. K., and Das, D. 2019 *Ceram. Inter.* **45** (18) 24835-24850.
[15] Hung,Y.W. and Ha, M. J. 2019 *J. Korean Ins. Electr. Electron. Mater. Eng.* **32** (4) 276-280.
[16] Sendi, R. 2019 *Moder. phys. Lett.* **33** (27) 1950336.
[17] Kulawik, J., Szwagierczak, D., and Skwarek, A. 2017 *Microelectron. Inter.* **40** (3) 158-162.
[18] Madani, F., Esnaashari, S. S., Webster, T. J., Khosravani, M. and Adabi, M. 2022 *JCR* **349** (1) 649-661.

Effect of Plasticisers and Zinc Oxide Nanoparticle Concentration on the Properties of Polyethylene Thin Films

Abeer Ahmed Alluhaibi*, Rabab Khalid Sendi

Umm Al- Qura University (UQU), Faculty of Applied Science, Physics Department, 21955, Makkah Al- Mukarramah, Saudi Arabia

* E-mail: s44286219@st.uqu.edu.sa

Abstract. The present work describes the preparation of a polyethylene film covered with a protective layer of zinc oxide nanoparticles (ZnO NPs). The presence of ZnO NPs on a slide supplemented with polyethylene was confirmed by the appearance of white cloud-like spots. Various ZnO NP concentrations were examined for their impact on optical and physical properties and in terms of inhibiting the growth of bacteria that can cause excessive irritation or intolerable skin sensitivity in the long term. This newly developed film holds potential for use in baby diapers to reduce bacterial contamination.

Keywords: Zinc oxide nanoparticles, nanopolymer, ultra-thin film, polyethylene

1. Introduction

Nanopolymer (NP) synthesis is an important part of polymeric nanotechnology because it involves the introduction of organic nanocomposites to improve the various properties of polymers. Thus, much research specializing in nanotechnology has focused on dispersing nanoscale metal oxides in the polymer matrix, as nanocomposites have similar advantages to polymers in terms of flexibility and the clarity of the film surface [1]. Nanomaterials are defined as particles with nanoscale dimensions, while NPs are extremely small particles in a catalytic reaction with thermal conductivity, chemical stability and nonlinear optical performance [2]. This is due to their high surface area per unit volume [3]. Moreover, nanoparticles have been integrated into many sectors of the medical, health, food, chemical and cosmetic industries, as well as some types of medicine, within which the manufacture of environmentally friendly nanomaterials is taken into account [4].

Researchers have shown interest in nanomaterials mixed with polymers. Given that nanostructured block copolymers in developed systems exhibit structuration at nanoscales and novel properties, synergistic effects may emerge, rendering such systems interesting and effective [5]. Inorganic nanoparticles exhibit distinctive optical and magnetic properties; they may also considerably enhance the hybrid's mechanical and physical properties. Some of the most prominent nanocomposites are zinc oxide (ZnO) NPs, which have received great interest since their discovery, as ZnO is known as an important electronic and photonic because of its semiconducting and piezoelectric properties [6,7]. It is classified as a semiconductor and is considered a biologically safe material with photo-oxidizing effects, as well as an excellent photocatalyst [8]. It has a 3.37 V band gap and a binding power of up to 60 megavolts. ZnO NPs also has a superior ability to absorb ultraviolet (UV) rays and a high exciton separation energy (~60 MeV); it also has chemical activity, optic and UV light-filtering properties, wound healing abilities and anti-inflammatory characteristics [9-15]. Because of its UV-filtering properties, it has been widely used in cosmetics, such as sunscreen lotions. Low-density polyethylene (LDPE) is a semi-crystalline polymer used in the production of insulating materials for electric power cables because of its unique mechanical and electrical properties [16]. Optical absorption spectra of all non-crystalline and crystalline materials reveal important information about the band structure and energy gap [17,18].

The results revealed the interaction efficiency of the thin films of polyethylene and ZnO nanoparticles, with the films appearing in the form of adhesives containing zinc nanoparticles (i.e. ZnO). According to a previous literature review, zinc compounds have effective antibacterial activity and affect the transparency and opacity of the films formed between them [20, 21]. Different proportions were used when a drip of LDPE was used to obtain a thin film, considering the results of LDPE stress resistance [22]. Filling an appropriate number of inorganic nanoparticles in a polymer matrix can enhance the tribological, mechanical and optical properties of the polymer, such as Zr [23], Si [24], CuS [25], ZnO [26], CuO and Ti [27]. Corrosion resistance in particles is due to some of the improved mechanical properties of the polymers, such as an increase in modulus and hardness [20]. ZnO has high strength and hardness, electrical conductivity and effective optical bacterial resistance [28-29]. In recent years, several studies have been conducted on polymer composites coated with nano-ZnO. Li et al. [15] explained the wear and friction of polytetrafluoroethylene with ZnO NPs in 15% of its total volume. Wang et al. [30] conducted various investigations on the tribological behavior of ZnO particles and composites made of nylon filled with filaments. They found that the use of ZnO filaments and ZnO particles resulted in an effective improvement in the performance of the tribological and physical properties of the nylon material. Subramani et al. [31] studied polypropylene filled with nano-ZnO and obtained results consistent with the previous study: ZnO demonstrated outstanding antibacterial activity against two types of human pathogenic bacteria, Klebsiella pneumoniae and Staphylococcus aureus. Lee et al. [32] also discovered improvements in the physical properties of thin linear LDPE, in addition to new antibacterial properties with the lowest concentration of ZnO NPs.

2. Experimental Design

2.1. Spin Coating

Spin coating is the simplest methodology for fabricating a thin film on a substrate. In this process, a drop of the additive is loaded into the center of a surface on the substrate, which is then turned at high speed. A repeated deposition is applied to extend the thickness of the thin films as desired and the substrate spins around an axis that ought to be perpendicular to the coating. As a result of centrifugal forces, the liquid spreads and a uniform liquid layer forms on the substrate. In general, an increase in the spin rate causes a decrease in film thickness. Spin rates of 3,000 rpm are accepted as a standard for chemical

IOP Conf. Series: Materials Science and Engineering 1269 (2022) 012002 doi:10.1088/1757-899X/1269/1/012002

compound applications. The spin-coating procedure includes acceleration steps, typically at a slower pace of 500 rpm per 5 s.

3. Materials and Methods

3.1. Materials

For the preparation of nanopolymer, sodium hydroxide (Merck KGaA, reagent grade, ≥98%, pellets), zinc diacetate (Merck, ACS reagent, ≥98%), and ethyl alcohol, pure (, 190 Proof, spectroscopic grade ACS, 95.0%). were used as reagents for modifications of the organic nanoparticles, with hexane as a solvent (JT Baker 99.8%). Low-density polyethylene (LDPE), the USP Reference Standard, was purchased from Merck.

3.2. *Sample Preparation*

3.2.1. ZnO Nanopowder Preparation

ZnO nanopowder was manufactured via the sol-gel technique, which included dissolution, hydrolysis, polymerization and conversion to ZnO. The formation contained 2 g of zinc acetate dihydrate

$Zn(CH_3COO)_2 \, . 2H_2O$. The powder was blended in distilled water (15 ml). Before ZnO was deposited and the nanopowder was obtained, the pH of the solution was adjusted between 9 and 12 using sodium hydroxide (NaOH). Thereafter, the solution was dried for several hours to obtain ZnO nanoparticles.

3.2.2. Thin Film Preparation

ZnO NP compounds were used to prepare the LDPE samples. First, appropriate amounts of LDPE were weighed and then drip-tilted onto ZnO nanocomposites according to the experimental conditions, with a precise balance between the concentration (1%-5%). The mixture was then transferred to a prepared relatively low-temperature annealing system, and the mixture was heated to 160°C and homogenized. Thereafter, the mixture consisting of ZnO and polyethylene was spread on a slide installed on the rotary coating device at a speed of up to 500–800 laps for thin film. The thin films obtained were annealed using an oven system at 120°C at a speed of 80 rpm. Using scissors, the thin films were cut into small pieces for analysis.

4.Results and Discussion

4.1. ZnO Nanoparticle Characterisation

The ZnO nanoparticles took the form of a hexagonal wurtzite structure (Figure 1). ZnO is a semiconductor with an energy bandgap of approximately 3.22 eV. As a result, ZnO accumulates UV-visible (Vis) light through an electronic excitement method linked to conduction and valence bands [32]

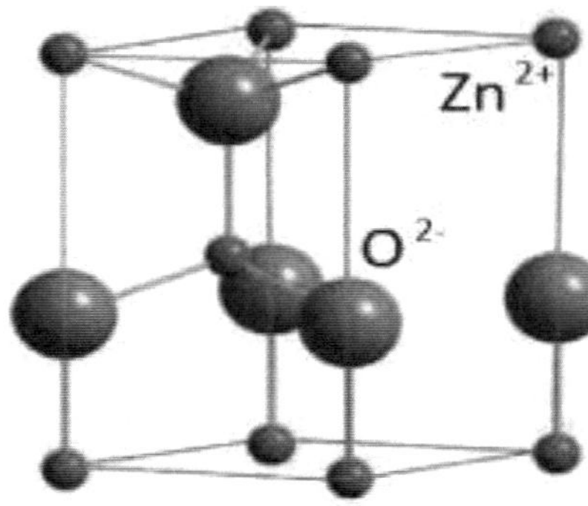

Figure 1. The wurtzite structure model of ZnO.

Equivalent outcomes have been observed for other ZnO-containing membranes [32, 33]. Previous experiments have found that ZnO NPs can slightly agglomerate within the deeper layers of material. The stress-strain curves were also used to calculate Young's modulus, strength and elongation at break. The addition of ZnO to LDPE thin films significantly alters their mechanical and optical properties. The action of the ZnO molecule as a device provides better strain at the fracture modulus regardless of the intensity of this additive due to 'surface adhesion and interaction between the ZnO NPs and LDPE' [33].

4.2. Scanning Electron Microscopy (SEM)

Scanning electron microscopy (SEM) is a type of electron microscopy that employs a beam of focused electrons of relatively low energy as an electron probe that is scanned over the specimen regularly (Thermo-Scientific Scios 2). The prepared films were physically stable and adhered very well to the LDPE substrates. After deposition, no cracking or peeling of the films was observed. Figure 2 depicts four SEM micrographs of LDPE/ZnO thin film that appeared as white dots on a grey background. The grain size increased as the thickness increased to 100 nm. Furthermore, for the thin film, the surface of the LDPE was visibly full of ZnO as the roughness of the pre-plasticized polymer's surface disappear.

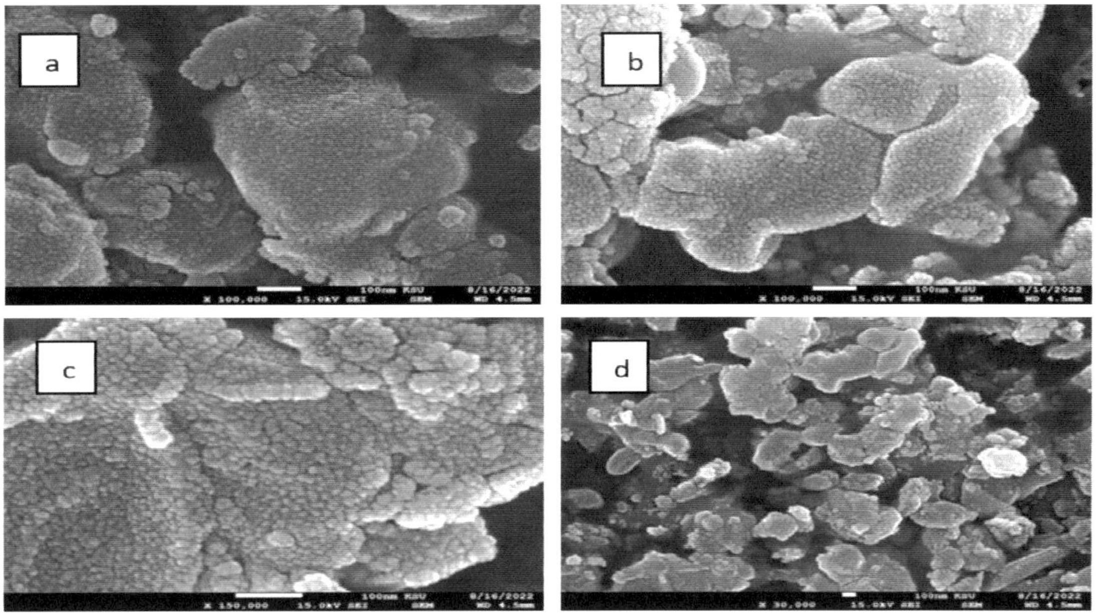

Figure 2. (a, b) white clouded clots with irregular distribution for ZnO/LDPE thin Film. (c, d) SEM micrographs for the surface of LDPE/ZnO for 100nm of LDPE/ZnO thin film

The spherical and composite ZnO nanoparticles that were exposed to different temperatures when forming the thin film were also clearly visible, as were the changes in the shape of the nanoparticles with LDPE. We conclude that with increasing nucleation rates, agglomeration appears in the sample due to the direct proportionality between nucleation and annealing rates.

Polyethylene has high-crystallinity structures, which means that to dissolve the polymer within the solvent, the solution must be heated above the compound melting temperature of $\approx$120°C for polyethylene. A solution of polyethylene in propanone and toluene is possible but requires 8 h to be obtained. When any solvent was used and the temperature was above 130°C, complete polymer dissolution was achieved in less than 4 h. When the temperature was below this value (i.e., at 80°C), the PE solution was non-uniform with its high viscosity, as illustrated in Figure 3 (a, b). The same applied to ZnO, which was annealed at different temperatures. The raised layer on the glass slide showed a surface dotted with small white dots. The surface of the thin film, annealed at 150°C, also contained some cracks, whereas no cracking was observed after annealing at 300°C. To achieve a thinner density, we increased the annealing temperature of the ZnO films from 150°C to 300°C. The size of the sphere-like points increased, and the films became more flexible. The cracks appeared smaller and more dispersed, resulting in a homogeneously shaped polymeric film.

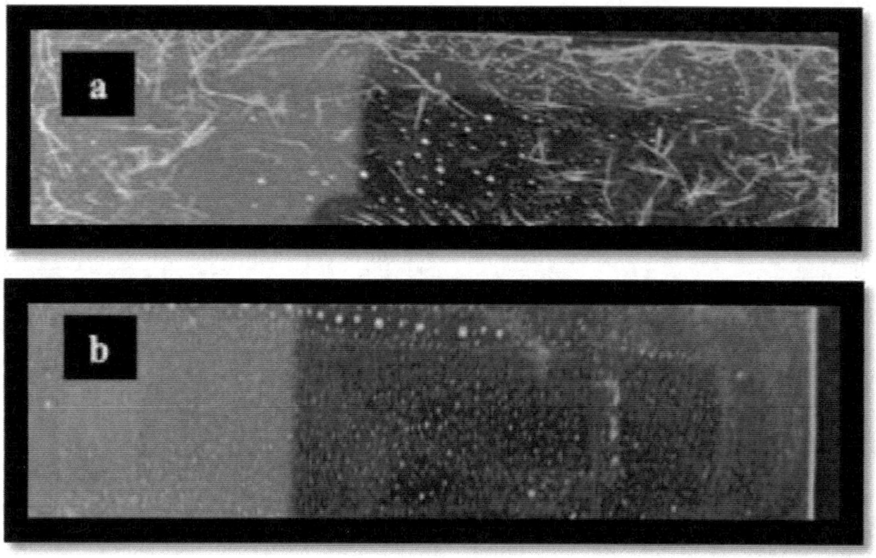

Figure 3. Polyethylene with ZnO films (a) on a cold at 80°C and (b) on a hot at 300°C.

Considering the mensuration conditions and experiment, the ZnO NP could be seen to partially agglomerate within the material's deeper layers (Figure 4), thus allowing for dispersion to occur in the polymer compound.

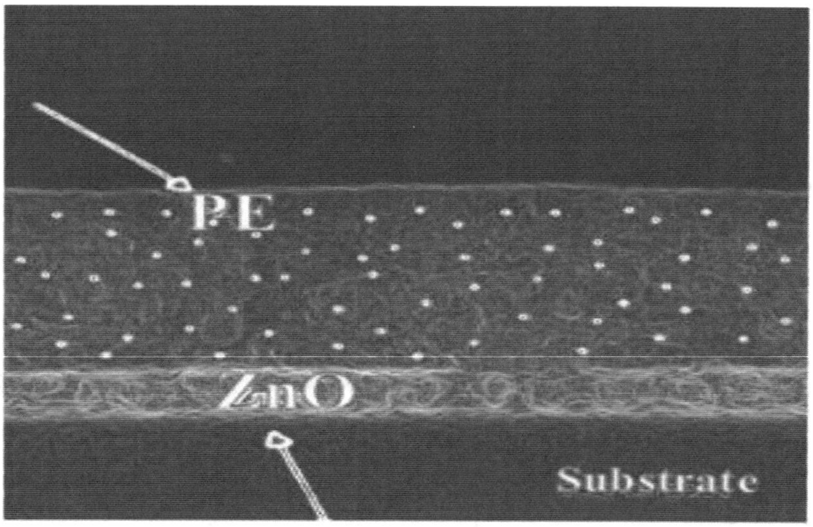

Figure 4. Schematic presentation of LDPE /ZnO thin film during the thermal annealing process.

4.3. Transmission Electron Microscopy (TEM)

ZnO nanoparticles are widely distributed in NPs, as demonstrated. However, when the mass fraction of ZnO nanoparticles was increased to 1% (Fig. 1a), slight agglomeration was observed, possibly due to the strong attraction of the nanoparticles. The TEM images of a 3% ZnO nanoparticle-containing NP

LDPE film show that the nanoparticles are evenly distributed throughout the NP, with aggregates ranging in size from 10 nm to 20 nm (Figs. 1c and 1d). The amount of agglomerates increased, and their volume became more irregular as the mass fraction of ZnO nanoparticles increased to 5% (Fig. 1d). This is well understood to be dependent on a number of NPs' morphological and chemical properties.

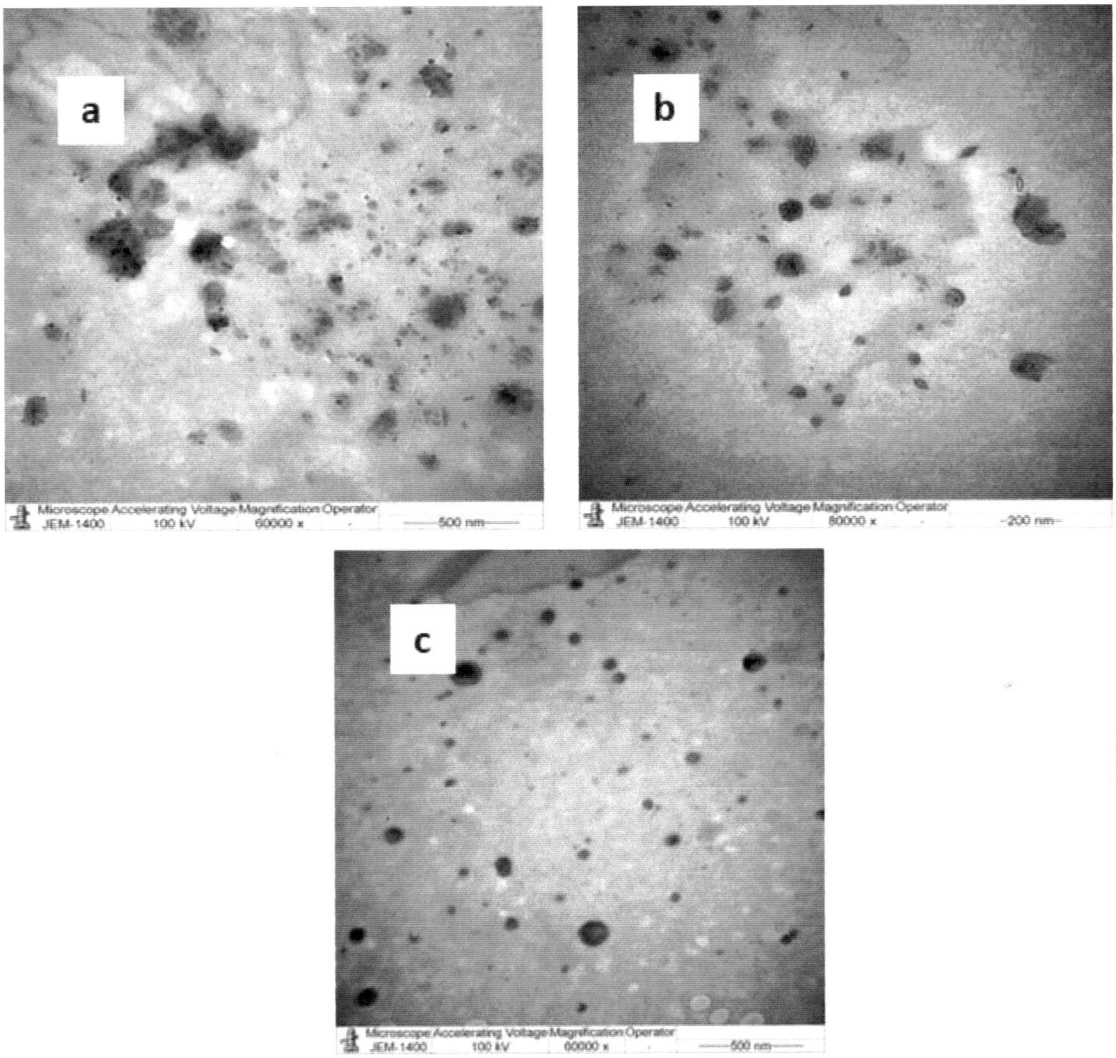

Figure 5. TEM images of thin film: a) LDPE films with 1 % ZnO nanoparticles, b) LDPE films with 3 % ZnO nanoparticles, c) LDPE films with 5 % ZnO nanoparticles

These images also revealed that the formed films contained spherical particles. Electron diffraction in the region selected for magnification also indicated the complexity of the crystal structure of the resulting thin films. Figure (6) also shows that the manufactured ZnO nanoparticles are not single crystals but rather appear as a group consisting of several single crystals.

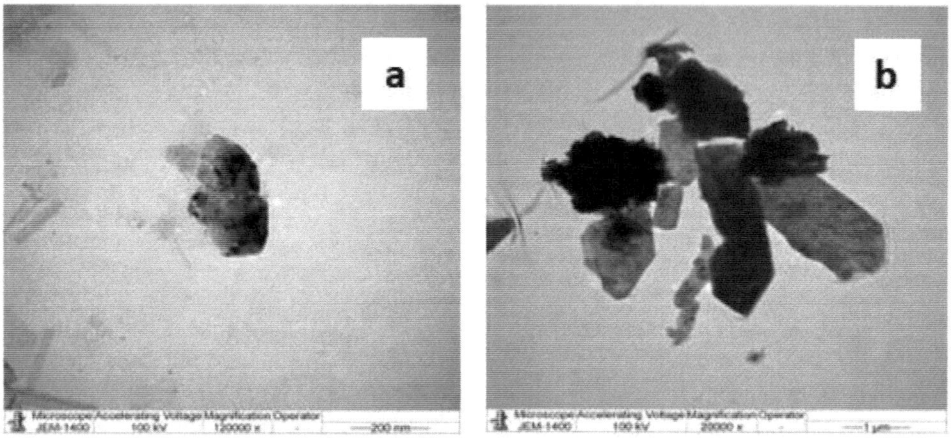

Figure 6. Shows the electron diffraction patterns of ZnO nanoparticles annealed via the TEM images (a) single crystal. (b)several crystals.

4.4. Fourier Transform Infrared Spectroscopy (IR)

An Absorption Spectra 3 Tri-Range FT-IR Spectrometer was used. Figure 7 illustrates the IR spectrum of the LDPE/ZnO NP synthesized using the spin-coating methodology, which was acquired in the 400–4000 cm^{-1}. The sharpness of the peaks increased with increasing annealing temperatures, which may lead to an increase in the crystalline structure of the LDPE/ZnO thin film perovskite structure of ZnO with increasing annealing temperatures. The IR also revealed that increasing the annealing temperature sharpens the diffraction peak for NP, inferring that the crystalline nature of ZnO increases with temperature. The peaks in the range of 896.54 –3616.11 cm^{-1} correspond to methylene (CH_2) groups. The adsorbed band at 1592.11-3616 cm^{-1} is assigned to O–H bending vibrations.

CH_2 groups correspond to the stretching modes at 2200 and 2400 cm^{-1} and deformations at 1423.92 and 2500 cm^{-1}(Figure 7). After annealing, unique peaks appeared at 3680.25 cm^{-1}, equivalent to C = O stretching vibration and at 3703.05 cm^{-1}, corresponding to hydroxyl group (–OH) vibrations. In contrast, the spectrum of the LDPE/ZnO thin film demonstrated a slight shift in the bands. The IR spectra for the NP composites, as illustrated in Figure 8, were consistent with this expectation. The effect of ZnO nanoparticles on the mechanical properties of the polymer is shown. From these results, we conclude that the tensile modulus is affected by the concentration of the polymer, which is approximately 20% lower than that of LDPE. Furthermore, the tensile modulus was affected by the level of reinforcement attained in the LDPE/ZnO NP. ZnO NPs consist of surfaces that tend to facilitate stronger slides, as they contain reactants that bond with polyethylene. On the other hand, Table 1 shows the intensity value of thin film.

Table 1. The results of LDPE/ZnO thin film using an IR Spectrometer.

	Peak	Intensity	Corr. Intensity	Base (H)	Base (L)	Area	Corr. Area
1	762.56	57.28	11.26	825.28	572.99	8536.583	1124.464
2	897.97	46.21	24.68	1275.69	825.28	15120.202	6179.509
3	1423.92	96.14	1.57	1435.32	1411.09	73.365	17.622
4	1592.11	93.98	2.47	1597.81	1576.43	86.364	24.537
5	3616.11	98.28	1.53	3631.79	3608.98	16.633	15.116
6	3641.76	98.49	1.41	3651.74	3631.79	15.190	13.159
7	3680.25	98.34	1.54	3687.37	3668.84	12.111	12.152
8	3703.05	96.90	3.95	3740.11	3687.37	0.266	94.704
9	3782.87	95.27	6.69	3815.65	3747.24	112.119	251.534

IOP Conf. Series: Materials Science and Engineering 1269 (2022) 012002 doi:10.1088/1757-899X/1269/1/012002

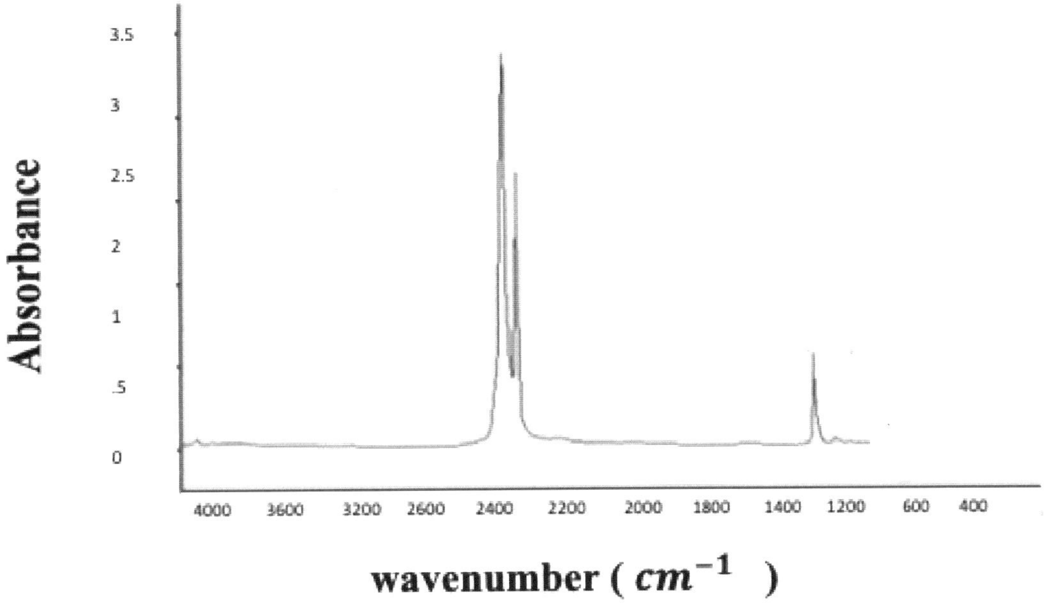

Figure 7. FTIR spectra of LDPE thin film to describe methylene (CH_2) groups in LDPE and increase of characteristic peaks for LDPE

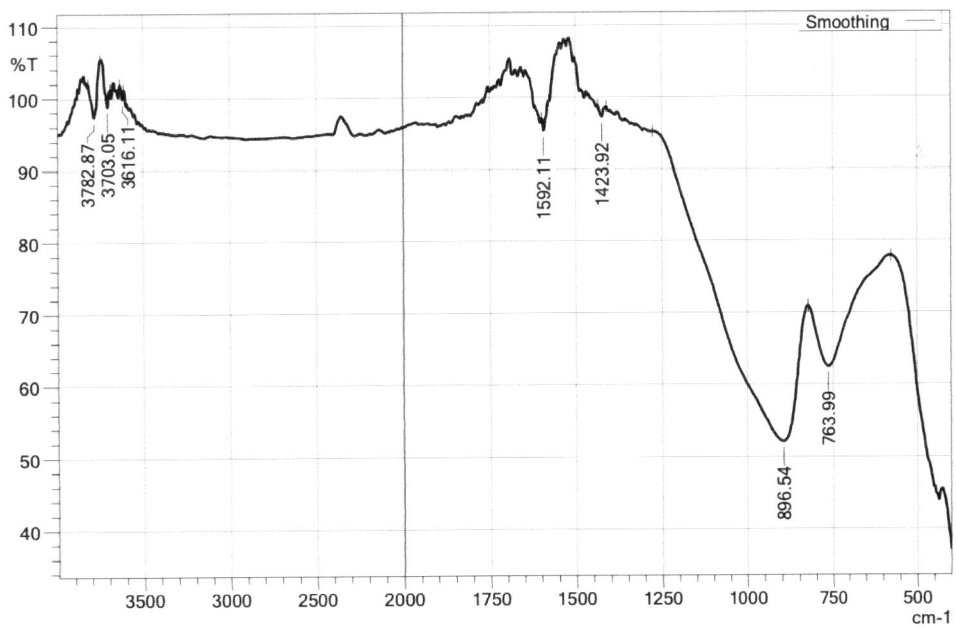

Figure 8. low-density Polyethylene with zinc oxide spectra (thin film form) by IR. Range correspondent to LDPE bands at 400–4000 cm^{-1}.

IOP Conf. Series: Materials Science and Engineering 1269 (2022) 012002 doi:10.1088/1757-899X/1269/1/012002

4.5. UV-Visible Spectroscopy

This absorption spectroscopy utilizes electromagnetic radiation ranging from 190 nm to 800 nm and is classified as UV (190–400 nm) and visible (Vis, 400–800 nm) (Thermo-Scientific Evolution 220). According to the absorption spectrum, as the particle size decreases, the band gap widens. An interaction effect also exists between the band gap and the wavelength of absorption.

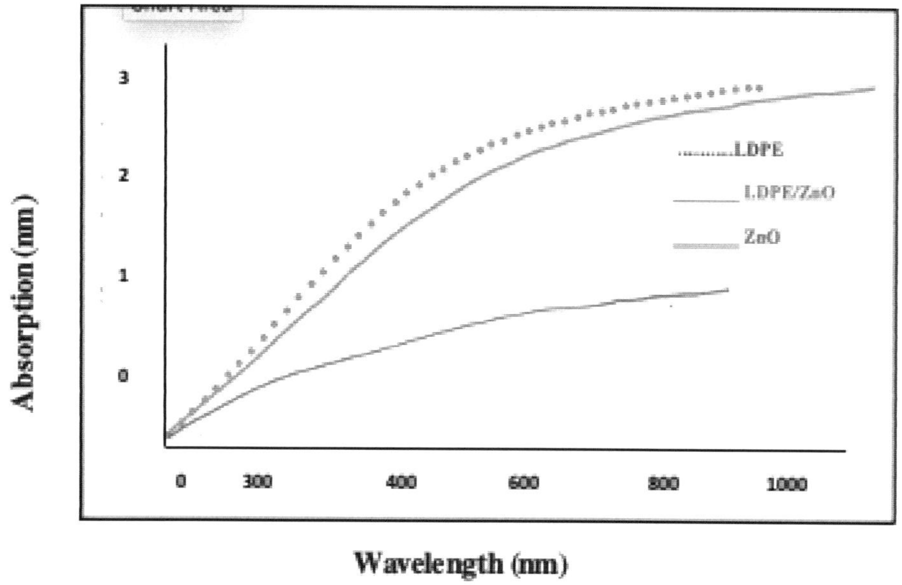

Figure 9. illustrates the transmittance spectra of the unfilled and ZnO-filled LDPE thin films

As previously stated, bulk ZnO absorbs at 385 nm [34]. The high blue shift absorption of synthesized

ZnO NPs over bulk ZnO could be attributed to significant particle size reduction. Figure 9 indicates that absorption was higher for the ZnO-filled LDPE samples when compared with the unfilled LDPE samples. The absorption spectra revealed that the chemical structure of LDPE was not modified by the ZnO NPs but instead remained in dispersed form. As the concentration of ZnO NPs increased, an increase in transmittance was observed. This transmittance percentage allowed the successful incorporation of ZnO into the ZnO/LDPE NP. As the results indicate, ethylene is clear within the case of the film; in contrast, it is opaque due to the presence of giant lumps, as well as an oversized variety of crystals within and the powerful lightweight scattering of light. This transmittance percentage indicates that the incorporation of ZnO in the ZnO/LDPE NP was successful.

5. Conclusion

In line with the hypothesis, we used materials to make ZnO in polyethylene films to produce a thin film that is water resistant and has a good moisture barrier. This caused synergistic effects because of their optical and mechanical properties. This study explained the integrated nanoparticles of their combination in polymer to organize the thin film. The SEM revealed a comparatively similar distribution of nanoparticles on the NP fracture surfaces. The thin film containing ZnO NPs displayed reduced UV light transmission, which might improve resistance to the harmful effects of UV light.

Acknowledgements

The authors is grateful to Umm Al-Qura University and the supervisor, Dr. Rabab Khalid Sendi, for the encouragement and support, and for extracting this review in an easy and useful way.

References

[1] Abbasi, A.R., et al., *Synthesis and characterization of Ag nanoparticles@ polyethylene fibers under ultrasound irradiation.* Ultrasonics sonochemistry, 2012. **19**(4): p. 853-857.
[2] Sondi, I. and B. Salopek-Sondi, *Silver nanoparticles as antimicrobial agent: a case study on E. coli as a model for Gram-negative bacteria.* Journal of colloid and interface science, 2004. **275**(1): p. 177-182.
[3] Harrasser, N., et al., *Antibacterial potency of different deposition methods of silver and copper containing diamond-like carbon coated polyethylene.* Biomaterials research, 2016. **20**(1): p. 1-10.
[4] Hoene, A., et al., *In vivo evaluation of copper release and acute local tissue reactions after implantation of copper-coated titanium implants in rats.* Biomedical Materials, 2013. **8**(3): p. 035009.
[5] Tiwari, V., et al., *Mechanism of anti-bacterial activity of zinc oxide nanoparticle against carbapenem-resistant Acinetobacter baumannii.* Frontiers in microbiology, 2018. **9**: p. 1218.
[6] Mania, S., et al., *The synergistic microbiological effects of industrial produced packaging polyethylene films incorporated with zinc nanoparticles.* Polymers, 2020. **12**(5): p. 1198.
[7] Espitia, P.J.P., et al., *Zinc oxide nanoparticles: synthesis, antimicrobial activity and food packaging applications.* Food and bioprocess technology, 2012. **5**(5): p. 1447-1464.
[8] Jones, N., et al., *Antibacterial activity of ZnO nanoparticle suspensions on a broad spectrum of microorganisms.* FEMS microbiology letters, 2008. **279**(1): p. 71-76.
[9] Chaudhry, Q., et al., *Applications and implications of nanotechnologies for the food sector.* Food additives and contaminants, 2008. **25**(3): p. 241-258.
[10] Bumbudsanpharoke, N., et al., *Zinc migration and its effect on the functionality of a low density polyethylene-ZnO nanocomposite film.* Food Packaging and Shelf Life, 2019. **20**: p. 100301.
[11] Rokbani, H., F. Daigle, and A. Ajji, *Long-and short-term antibacterial properties of low-density polyethylene-based films coated with zinc oxide nanoparticles for potential use in food packaging.* Journal of Plastic Film & Sheeting, 2019. **35**(2): p. 117-134.
[12] Al-Naamani, L., J. Dutta, and S. Dobretsov, *Nanocomposite zinc oxide-chitosan coatings on polyethylene films for extending storage life of okra (Abelmoschus esculentus).* Nanomaterials, 2018. **8**(7): p. 479.
[13] Al-Tayyar, N.A., A.M. Youssef, and R. Al-Hindi, *Antimicrobial food packaging based on sustainable Bio-based materials for reducing foodborne Pathogens: A review.* Food chemistry, 2020. **310**: p. 125915.
[14] Tharayil, A., S. Banerjee, and K.K. Kar, *Dynamic mechanical properties of zinc oxide reinforced linear low density polyethylene composites.* Materials Research Express, 2019. **6**(5): p. 055301.
[15] Rojas, K., et al., *Effective antimicrobial materials based on low-density polyethylene (LDPE) with zinc oxide (ZnO) nanoparticles.* Composites Part B: Engineering, 2019. **172**: p. 173-178.
[16] Klapiszewski, Ł., et al., *A high-density polyethylene container based on ZnO/lignin dual fillers with potential antimicrobial activity.* Polymer Testing, 2019. **73**: p. 51-59.

[17] Khoirunnisa, A.R., et al. *UV-screening, transparency and water barrier properties of semi refined iota carrageenan packaging film incorporated with ZnO nanoparticles*. in *AIP conference proceedings*. 2018. AIP Publishing LLC.
[18] Castro-Mayorga, J.L., et al., *The impact of zinc oxide particle morphology as an antimicrobial and when incorporated in poly (3-hydroxybutyrate-co-3-hydroxyvalerate) films for food packaging and food contact surfaces applications.* Food and Bioproducts Processing, 2017. **101**: p. 32-44.
[19] Wei, L., T. Tang, and B. Huang, *Synthesis and characterization of polyethylene/clay–silica nanocomposites: a montmorillonite/silica-hybrid-supported catalyst and in situ polymerization.* Journal of Polymer Science Part A: Polymer Chemistry, 2004. **42**(4): p. 941-949
[20] Yañez, D., et al., *Photocatalytic inhibition of bacteria by TiO2 nanotubes-doped polyethylene composites.* Applied Catalysis A: General, 2015. **489**: p. 255-261.
[21] Fang, Y., et al., *Properties of mLLDPE/LDPE blends in film blowing.* Polymer Engineering & Science, 2005. **45**(3): p. 343-353.
[22] Wang, Q., et al., *The effect of particle size of nanometer ZrO2 on the tribological behaviour of PEEK.* Wear, 1996. **198**(1-2): p. 216-219.
[23] Wang, Q., Q. Xue, and W. Shen, *The friction and wear properties of nanometre SiO2 filled polyetheretherketone.* Tribology international, 1997. **30**(3): p. 193-19
[24] Zhang, H.-J., et al., *Surface modification of CuS nanoparticles and their effect on the tribological properties of hybrid PTFE/kevlar fabric/phenolic composite.* Journal of composite materials, 2010.**44**(21): p. 2461-2472.
[25] Li, F., et al., *The friction and wear characteristics of nanometer ZnO filled polytetrafluoroethylene.* Wear, 2001. **249**(10-11): p. 877-882.
[26] Bahadur, S. and C. Sunkara, *Effect of transfer film structure, composition and bonding on the tribological behavior of polyphenylene sulfide filled with nano particles of TiO2, ZnO, CuO and SiC.* Wear, 2005. **258**(9): p. 1411-1421.
[27] Demjén, Z., B. Pukánszky, and J. Nagy, *Evaluation of interfacial interaction in polypropylene/surface treated CaCO3 composites.* Composites Part A: Applied Science and Manufacturing, 1998. **29**(3): p. 323-329.
[28] Zhang, L., et al., *ZnO nanofluids–A potential antibacterial agent.* Progress in Natural Science, 2008. **18**(8): p. 939-944.
[29] Liu, Y.-j., et al., *Antibacterial activities of zinc oxide nanoparticles against Escherichia coli O157: H7.* Journal of applied microbiology, 2009. **107**(4): p. 1193-1201.
[30] Li, S.C. and Y.N. Li, *Mechanical and antibacterial properties of modified nano-ZnO/high-density polyethylene composite films with a low doped content of nano-ZnO.* Journal of applied polymer science, 2010. **116**(5): p. 2965-2969.
[31] Wang, S., S. Ge, and D. Zhang, *Comparison of tribological behavior of nylon composites filled with zinc oxide particles and whiskers.* Wear, 2009. **266**(1-2): p. 248-254.
[32] Chandramouleeswaran, S., et al., *Functional behaviour of polypropylene/ZnO–soluble starch nanocomposites.* Nanotechnology, 2007. **18**(38): p. 385702.
[33] Emamifar, A. and Mohammadizadeh, M.*Preparation and application of LDPE/ZnO nanocomposites for extending shelf life of fresh strawberries.* Food technology and biotechnology, 2015. **53**(4): p. 488-495.
[34] Zvyagin, A.V., et al., *Imaging of zinc oxide nanoparticle penetration in human skin in vitro and in vivo.* Journal of biomedical optics, 2008. **13**(6): p. 064031.

IOP Conf. Series: Materials Science and Engineering 1269 (2022) 012003 doi:10.1088/1757-899X/1269/1/012003

Correlating stimulated emission phase to the gain spectra in semiconductor lasers

Abanoub Mikhail[1], Safwat William Zaki Mahmoud[2]

[1,2]Faculty of Science, Minia University, Minia, Egypt

Email: abanoub.mikhail@mu.ed.eg , safwatzaki@mu.edu.eg

Abstract. In laser physics, the incident electric field and the stimulated field are assumed to have the same frequency, direction of propagation, polarization, and phase —same state. However, no formal proof of the phase identity (zero phase shift) was reported. The existing theories on the phase shift are in conflict with each other and with the phase identity. The phase shift in three semiconductor active media (GaAs, $In_{0.15}Ga_{0.85}As$, and $In_{0.48}Ga_{0.58}As_{0.9}P_{0.1}$) has been studied using a semiclassical approach. Contrary to the conventional belief, it has been found that the phase shift is not zero. The phase shift is not even a single value but rather a spectrum corresponding to the gain spectrum. At a carrier concentration of $2.5 \times 10^{24}\ m^{-3}$, the minima of the phase shift spectra are 1.5597, 1.509, and 1.399 radians for GaAs, $In_{0.15}Ga_{0.85}As$, and $In_{0.48}Ga_{0.58}As_{0.9}P_{0.1}$, respectively. Stimulated emission is shown to occur whenever the phase shift is positive and lies in the interval between 0 and π/2. Because of radiation reaction of the radiating source and the finite lifetime of excited states, stimulated emission cannot attain the same phase as the incident field —no zero-phase shift. These results reveal that the conventional picture of phase identity is incorrect, and phase matching between the incident field and stimulated is not a necessary condition for stimulated radiation. Despite the fundamentality of the concept of stimulated emission phase, such outcomes are not reported nor discussed in the literature.

1. Introduction

The process of stimulated emission was predicted by Einstein based on mechanical and thermodynamical arguments in 1917 [1]. He stated that the energy exchange between the incident electromagnetic field and Plank oscillators could be positive or negative depending on the phases of the incident radiation field and the oscillators. The problem of the phase shift (phase difference) between the incident radiation field and the stimulated field was not dealt with in Einstein's work.

When an electromagnetic wave passes through a medium, the wave creates oscillating dipoles. These dipoles initially absorb energy. If the electromagnetic wave continues to propagate in the medium, the oscillating dipoles start to radiate energy in a way that is temporary and spatially coherent with the driving field —stimulated emission [2]. Determining the phase shift between the incident electric field and the stimulated field is complicated for multiple reasons. Firstly, the radiation pattern of a system in free space is substantially altered when the same system is embedded in media. For example, in free space, the field lines of energy flow run radially outwards except for the dipole axis.

In a material medium, the energy flow lines are perpendicular to the dipole axis due to damping [3]. Secondly, the experimentally measurable physical quantity is not the phase shift but rather the intensity of radiation. Therefore, during conventional experiments, the phase shift information is lost. Nonetheless, Josef Kröll and coworkers managed to measure the phase and amplitude of the emitted field and correlated it with the input signal [4]. Their phase measurement experiment showed that the behavior of the phase at resonance gain is inverted with respect to resonance absorption.

The widespread view is that stimulated emission occurs in phase with the incident field. This point of view is taught in our universities without any mathematical or experimental proof. Our objective is to address this issue based on practical lasing media.

Pollnau argued that, at resonance, polarization leads the incident field by $\pi/2$ and the electric field scales with the distance between the two charges of the dipole. Therefore, stimulated emission must lead the incident field by $\pi/2$. According to Pollnau, phase shift of $\pi/2$ is a must for energy conservation to hold true [5].

Zhao and coworkers studied radiation emitted from a sheet of charge [6]. They found that the emitted field is delayed by $\pi/2$ from the incident counterpart. Being inconsistent with practical optical gain process, the authors proposed that if the radiated electric field from a single dipole is $\pi/2$ advanced in phase, then the superposition of the radiated field from the infinite sheet of dipoles would be in phase with the incident field. In the discussion section of this paper, we will show the reasons why Pollnau and Zhao et al. analyses are not sound.

Charles Henry found a relation between the phase change due to carrier injection and the gain peak [7]. However, the phase shift between the incident and the stimulated field due to the detuning frequency, at constant carrier density, was not dealt with.

In this work, the phase shift in three typical gain media GaAs, $In_{0.15}Ga_{0.85}As$, and $In_{0.48}Ga_{0.58}As_{0.9}P_{0.1}$ is investigated. These gain materials were chosen for their wide use in diode laser fabrication and in photonic devices, but other candidates could have been used instead. In addition to showing the fallacies in the work of Pollnau and Zhao et al., we illustrate why the conventional belief of phase identity cannot be true.

2. Method

Within the semi-classical model of electromagnetic waves interacting with semiconductors, the field is treated classically while the semiconductor is treated using the techniques of quantum statistical mechanics such as the density matrix. An incident electric field $\boldsymbol{E}_i(z,t) = \hat{x}E_z e^{i(\omega t - kz)}$ oscillates at a frequency ω and propagates in the z-direction with a wave number k. The field displaces the center of the positive nucleus and negative electron cloud from their equilibrium position forming oscillating dipoles. The medium polarization is related to the incident field by the constitutive relation [8]

$$\boldsymbol{P}(\omega) = \varepsilon_0 \chi \boldsymbol{E}_i(\omega) \tag{1}$$

where χ and ε_0 are the susceptibility and vacuum permittivity in order. The radiated electric field (E_r) from an active medium of thickness Δz is related to the medium polarization P [9]

$$E_r = -\frac{i}{2n_b}\sqrt{\frac{\mu_0}{\varepsilon_0}}\,\omega P \Delta z \tag{2}$$

where μ_0 and n_b are the vacuum permeability and host medium background refractive index, respectively. Within the free carrier theory and under the parabolic band approximation, the real and imaginary parts of the susceptibility are obtained as [10, 11]

$$\mathrm{Im}[\chi] = \frac{1}{\varepsilon_0}\int_{E_g}^{\infty} \frac{|\mu_{ba}|^2 g_{cv}(E_{ba})[f_c(E_{ba}) - f_v(E_{ba})]}{(\hbar\omega - E_{ba})^2 + (\hbar/\tau_d)^2} dE_{ba} \tag{3}$$

$$Re[\chi] = \frac{1}{\varepsilon_0}\int_{E_g}^{\infty} \frac{|\mu_{ba}|^2 g_{cv}(E_{ba})[f_c(E_{ba}) - f_v(E_{ba})](\hbar\omega - E_{ba})}{(\hbar\omega - E_{ba})^2 + (\hbar/\tau_d)^2} dE_{ba} \tag{4}$$

where E_g is the band gap energy, E_{ba} is the transition energy, τ_d is relaxation time, $g_{cv}(E_{ba})$ is the joint density of states, μ_{ba} is the dipole moment, $f_c(E_{ba})$ and $f_v(E_{ba})$ are the quasifermi distribution of electrons in the conduction and valence bands respectively. Equation (2) shows that the radiated field is always perpendicular to the polarization —the radiated field leads the polarization in phase by $\frac{\pi}{2}$ [9].

The gain coefficient is the fractional increase in the intensity of radiation per unit distance travelled in the medium. There is gain in the lasing medium only if $f_c(E_{ba}) > f_v(E_{ba})$. This condition limits the imaginary part of the susceptibility to the upper half of the complex plane. Thus, the phase shift between the stimulated field and the incident field becomes

$$\Delta\theta(\omega) = \frac{\pi}{2} - \tan^{-1}\left(\frac{n_b c}{\omega}\frac{g(\omega)}{Re[\chi]}\right) \tag{5}$$

where the gain coefficient $g(\omega)$ is given by [10]

$$g(\omega) = \frac{\omega}{n_b c} Im[\chi] \tag{6}$$

The equation for the phase shift (Eq. (5)) must be solved numerically for the gain medium under study. We have chosen GaAs, $In_{0.15}Ga_{0.85}As$, and $In_{0.48}Ga_{0.58}As_{0.9}P_{0.1}$ because these are common direct band-gap semiconductor materials. They emit laser light around 850, 1300 and 1550 nm, respectively.

The quasi-Fermi levels of the valence and conduction bands are numerically found using the charge neutrality condition imposed on the gain medium. Charge neutrality requires the electrons excited into the conduction band to leave an equal number of empty states behind in the valance band [12]. Afterwards, equations (3-6) are solved numerically to obtain the phase shift of the stimulated radiation as a function of frequency. Wolfram Mathematica (https://www.wolfram.com/mathematica/) has been used to carry out the calculations using the parameters in Table (1).

Parameter name	**Symbol**	**GaAs**	**$In_{0.15}Ga_{0.85}As$[a]**	**$In_{0.58}Ga_{0.42}As_{0.9}P_{0.1}$[b]**	**Units**
Band gap at 0K	$E_g(0)$	*1.519*	1.3031	0.84	*eV*
Band gap offset	Δ	0.34	0.062	0.326	*eV*
Electron effective mass	m_c/m_0	0.08	0.05569	0.0449	--
Heavy hole effective mass	m_v/m_0	0.5	0.4865	0.438	--
Refractive index	n_r	3.6	3.3948	3.514	--
Relaxation time	τ_d	10^{-13}	10^{-13}	10^{-13}	s

[a]extracted from [13] [b]extracted from [14]

3. Results

Figures (1-3) show the gain spectrum and the phase shift spectrum for the three materials under study at the threshold gain concentration. The main observations are:

Firstly, the phase shift for all the investigated materials is a spectrum corresponding to the gain spectrum. Secondly, the phase shift spectrum is always positive and less than $\frac{\pi}{2}$. Thirdly, stimulated radiation is produced by both resonant electromagnetic fields and quasi-resonance fields. However, the phase shift at a quasi-resonance incident field is larger than phase shift in case of resonant electromagnetic field. Fourthly, in the absorption region, the phase shift becomes negative. Taken together, the four observations show that the phase shift is a spectrum whose minimum is not at zero. This conclusion is true independent of the injected carrier concentration and attains validity for the three materials under study.

Figure (4) illustrates the phase shift minimum and the gain peak plotted against the carrier concentration. While the gain peak increases with carrier concentration, the minimum of the phase shift spectra decreases. At any fixed carrier concentration, the phase shift minimum of $In_{0.48}Ga_{0.58}As_{0.9}P_{0.1}$ is smaller other two materials. For example, at a carrier concentration of $\mathbf{2.5 \times 10^{24}\ m^{-3}}$, the phase shift spectrum minimum is $\mathbf{1.5597}$, $\mathbf{1.509}$, and $\mathbf{1.399}$ radians for GaAs, $In_{0.15}Ga_{0.85}As$ and $In_{0.48}Ga_{0.58}As_{0.9}P_{0.1}$, respectively. The reason for the low minimum of phase shift spectrum in case of $In_{0.48}Ga_{0.58}As_{0.9}P_{0.1}$ may be attributed to the smaller effective mass of the charge carriers in $In_{0.48}Ga_{0.58}As_{0.9}P_{0.1}$ as compared with the other two materials.

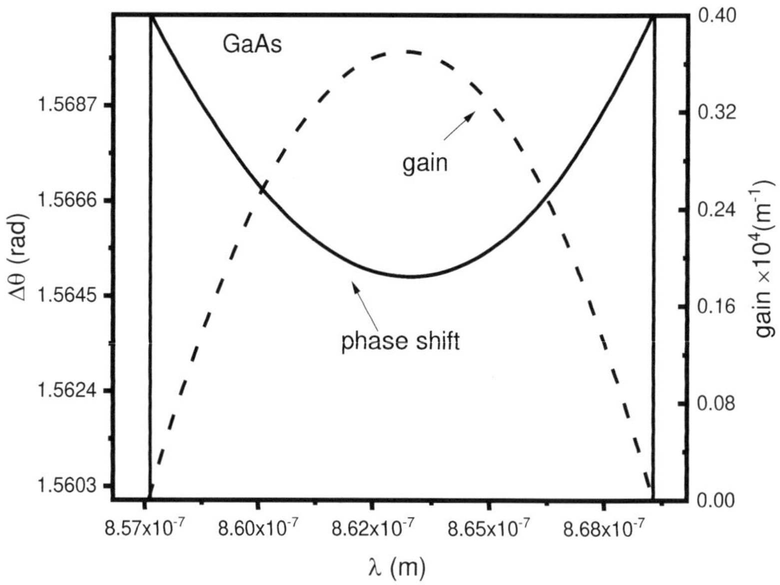

Fig. 1: Phase shift (solid) and gain spectrum of GaAs (dashed) at lowest gain carrier concentration of $\mathbf{2.38 \times 10^{24}\ m^{-3}}$ and temperature of 300K

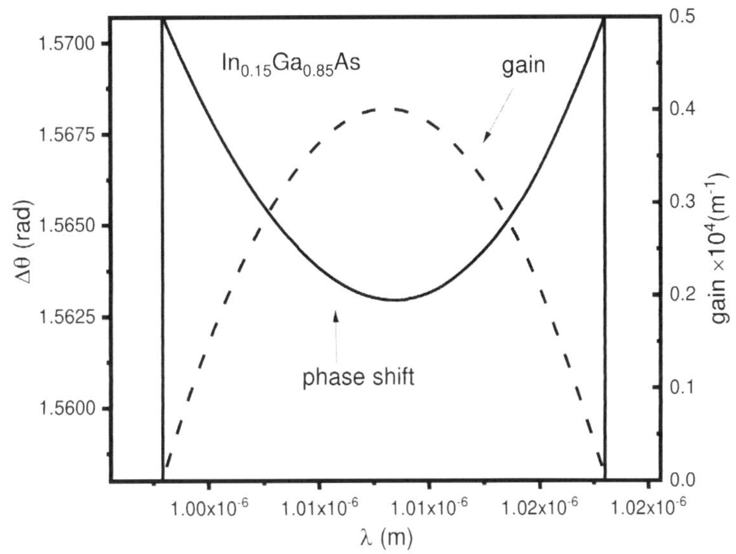

Fig. 2: Phase shift (solid) and gain spectrum of $In_{0.15}Ga_{0.85}As$ (dashed) at lowest gain carrier concentration of $1.74 \times 10^{24}\ m^{-3}$ and temperature of 300K

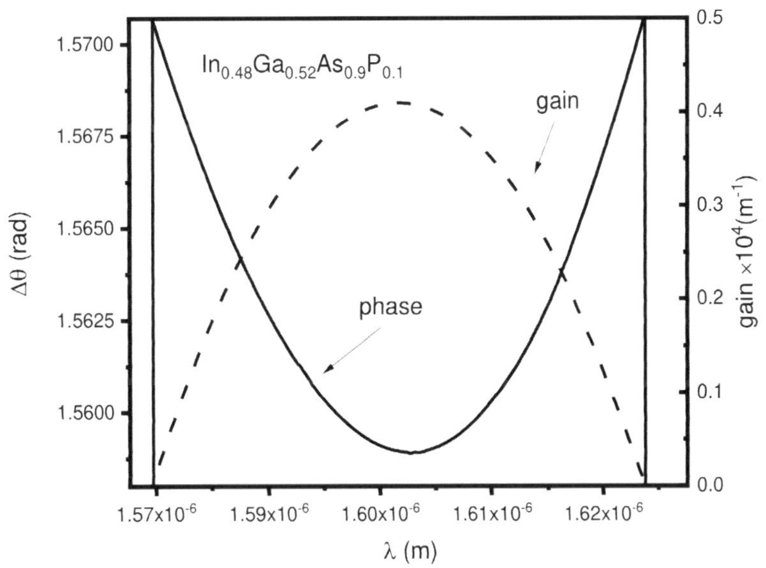

Fig. 3: Phase shift (solid) and gain spectrum of $In_{0.48}Ga_{0.58}As_{0.9}P_{0.1}$ (dashed) at lowest gain carrier concentration of $1.29 \times 10^{24}\ m^{-3}$ and temperature of 300K

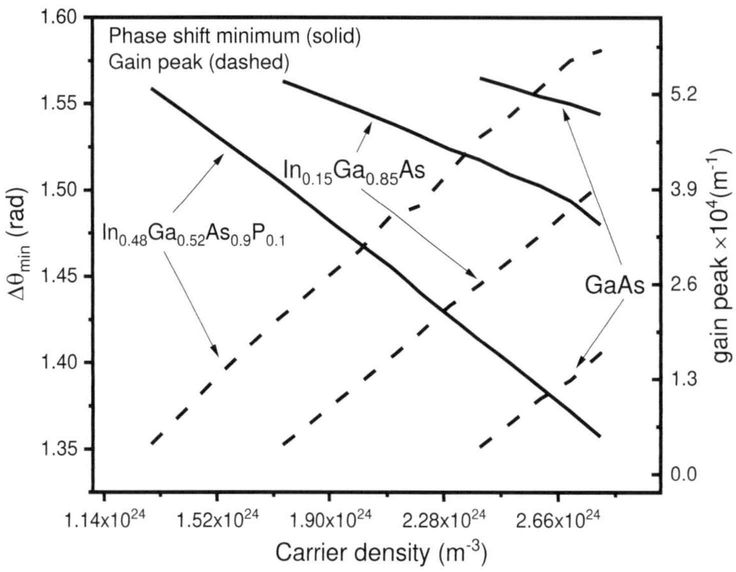

Fig.4: Phase shift minimum (solid), and gain peak (dashed) versus carrier concentration for the three materials under study at temperature of 300K

4. Discussion

The problem of stimulated emission phase is intimately linked to foundational concepts in physics. Stimulated emission results from the interference of the incident field with the field radiated from the atom [15]. Scattering and interaction between carriers could destroy the phase shift. Despite of adopting the free carrier theory, which neglects scattering and coulomb interaction between carriers, the phase shift turns out to attain nonzero values. There is a minimum in the phase shift spectrum, but its value is greater than zero. We believe there are compelling reasons that inhibit the phase shift from vanishing.

The first reason is the finite lifetime of excited states. Typically, GaAs-based-heterostructure semiconductors have carrier lifetime around 10^{-9} s [16, 17]. This means that it takes the stimulated photon time equal to the lifetime of the state to get fully emitted, but, by then, the phase of the emitted radiation would be different from the phase of the incident radiation. This argument can be extended to low dimension devices such as quantum dot lasers, and small lasers [16, 18, 19].

The second reason is the extended nature of atoms. Stimulated radiation cannot instantly detach itself from the radiating source. The incident field modulates the source's charge and current distributions at first. In the region surrounding the atoms (near field region), the stimulated electric and magnetic fields are attached to the current source. It is only after travelling a large distance compared to the source size that they become completely detached from the radiating source. Moreover, electromagnetic retardation may be another factor in delaying the emitted field. Retardation is a result of the finite speed of light. The state of radiation at time t reflects the state of the source of radiation at a prior time $t_0 = t - r/c$, where r is the distance travelled by the stimulated radiation [8].

The third and most important reason is radiation damping. The emitted radiation exerts a back reaction on the atom itself which has a friction-like damping effect [20]. Even if one managed to remove other dephasing mechanisms such as scattering of charge carriers. The stimulated radiation of electromagnetic waves undergoes a form of damping which is inherent to the radiation process and cannot be eliminated [21, 22].

What was the flaw in Pollnau's thinking? Let us first rephrase his argument. The number of photons Φ is proportional to the intensity of the electromagnetic wave I of a particular optical mode. If $\vec{E}_i$ and $\vec{E}_r$ represent the amplitude of the incident and radiated fields at a particular frequency, then

$$\Phi \propto \mathrm{I} = \frac{c\varepsilon_0}{2}\left|\vec{E}_i + \vec{E}_r\right|^2 = \frac{c\varepsilon_0}{2}\left[\left|\vec{E}_i\right|^2 + \left|\vec{E}_r\right|^2 + 2\left|\vec{E}_i\right|\left|\vec{E}_r\right|\cos(\Delta\theta)\right] \tag{7}$$

Prior to Pollnau's work, Melissa cray et al. showed that the first term in Eq. (7) is the energy of the incident field, the second term is the energy associated with the radiated field due to scattering and spontaneous emission, and the third term represents the interaction energy due to stimulated emission [15]. Pollnau main point is that unless the interference term (the 3rd term in Eq. (7)) vanishes, the conservation of energy principle is violated [5]. Therefore, he concluded that the phase shift must be $\frac{\pi}{2}$. However, this conclusion is incorrect because electric dipoles radiate waves in a manner such that the energy lost in one direction could be gained in other directions. This latter idea is supported by W. J. Cocke who studied the interaction of an incident plane wave with a charge distribution and found that the interaction term due to stimulated emission gains a net contribution exactly in the forward-scattering direction [23].

There exist flaws in the analysis of Zhao and coworkers—as hinted at in the introduction section of this paper. For a start, the authors did not provide any proof for the assumption that stimulated emission from a single atom (dipole) is $\pi/2$ advanced in phase with respect to the incident field [6]. Their approach seems to be an ad hoc attempt to make stimulated emission from an ensemble of atoms agrees with the conventional view of phase identity. In addition to that, no damping mechanism which causes transients to decay was included in their model. In fact, Einstein argued that stimulated radiation is a directed process. Thus, the recoil of the atoms must exist [1].

5. Conclusion

An investigation of the phase shift of stimulated emission of radiation and its relation to the gain in various lasing media was carried out. The phase shift of stimulated radiation is found to be a spectrum that follows the gain spectral profile. The gain spectrum and phase shift spectrum are related to each other. The optical gain is greater than zero only when the phase shift is in the interval $0 < \Delta\theta < \pi/2$. Therefore, stimulated emission is concluded to be possible over this range only.

The minimum of the phase shift spectrum seems to be a characteristic feature of any gain material. Generally, the smaller the minimum of the phase shift spectrum is, the higher is the gain peak. This observation suggests that the most fundamental quantity in classifying lasing materials is the phase shift. In such a scheme, the best lasing material that could ever be found is the one with the lowest possible phase shift minimum.

We have also elucidated the reasons for the unfeasibility of the conventional belief of zero stimulated emission phase shift. The causes lie in the finite lifetime of the excited states, the extended nature of the radiating sources, and most importantly radiation reaction. Future work in this direction of research may include the dependence of the phase shift on the temperature, interaction between charge carriers, and the study of the phase shift in the near field and in nanodevices.

References

[1] Einstein A 1917 The Quantum Theory of Radiation *Phys. Z.* **18**
[2] Siegman A E 1986 Lasers (Mill Valey, California: University Science Books)
[3] Li X, Pierce D M and Arnoldus H F 2011 Redistribution of energy flow in a material due to damping *Opt. Lett.* **36**: p. 349-351

[4] Kröll J, Darmo J, Dhillon S S, Marcadet X, Calligaro M, Sirtori C and Unterrainer K 2007 Phase-resolved measurements of stimulated emission in a laser *Nature* **449**: p. 698–701
[5] Pollnau M 2018 Phase aspect in photon emission and absorption *Optica* **5**: p. 465–474
[6] Zhao Z, Kong M and Xu Y 2016 Phase advance of photons generated in stimulated emission relative to incident light *Optik* **127** 8970-8975
[7] Henry C 1982 Theory of the linewidth of semiconductor lasers *IEEE J. Quantum Electron.* **18**: p. 259-264
[8] Jackson J D 1999 Classical electrodynamics (Singapore: John Wiley & Sons)
[9] Yamada M 2014 Theory of semiconductor lasers (Atlanta, USA: Springer)
[10] Suhara T 2004 Semiconductor laser fundamentals (New York: CRC press)
[11] Yamada M and Ishiguro H 1981 Gain calculation of undoped GaAs injection laser taking account of electronic intra-band relaxation *Jpn. J. Appl. Phys.* **20**: p. 1279-1279
[12] Blood P 2015 Quantum Confined Laser Devices: Optical gain and recombination in semiconductors OUP Oxford)
[13] Holly C, Hengesbach S, Traub M and Hoffmann D 2013 Simulation of spectral stabilization of high-power broad-area edge emitting semiconductor lasers *Opt. Express* **21**: p. 15553–15567
[14] Goldberg Y A and Schmidt N M 1999 Handbook series on semiconductor parameters (London: World Scientific)
[15] Melissa Cray M-L S and Milonni P W 1982 Stimulated emission, absorption and interference *Am. J. Phys.*
[16] Raymond S, Fafard S, Poole P, Wojs A, Hawrylak P, Gould C, Sachrajda S, Charbonneau S, Leonard D, Leon R J S and microstructures 1997 State-filling and magneto-photoluminescence of excited states in InGaAs/GaAs self-assembled quantum dots *Superlattices Microstruct.* **21**: p. 541-558
[17] Levinshtein M, Rumyantsev S and Shur M 1999 Ternary and quaternary III-V compounds
[18] Geiregat P, Van Thourhout D and Hens Z J N A M 2019 A bright future for colloidal quantum dot lasers *NPG Asia Mater.* **11**: p. 1-8
[19] Tiguntseva E, Koshelev K, Furasova A, Tonkaev P, Mikhailovskii V, Ushakova E V, Baranov D G, Shegai T, Zakhidov A A and Kivshar Y J A n 2020 Room-temperature lasing from Mie-resonant nonplasmonic nanoparticles *ACS Nano* **14**: p. 8149-8156
[20] Landau L D 2013 The classical theory of fields (Oxford, England: Elsevier)
[21] Wistisen T N, Di Piazza A, Knudsen H V and Uggerhøj U I J N c 2018 Experimental evidence of quantum radiation reaction in aligned crystals *Nat. Commun.* **9**: p. 1-6
[22] Macchi A 2018 Intense laser sheds light on radiation reaction *Physics* **11**: p. 13-13
[23] Cocke W J 1978 Stimulated emission and absorption in classical systems *Phys. Rev. A* **17**: p. 1713-1720

Evaluation of the seismic performance of lightweight concrete multistory buildings

Yasser A.S Gamal[1] and Mostafa Abd Elrazek[2]
[1] Lectures of Structural, Civil Eng. Department, Higher Institute of Engineering and Technology, New Minia, Egypt
[2] Lectures of properties and strength of materials, Civil Eng. Department, Higher Institute of Engineering and Technology, New Minia, Egypt

E-mail: (Yasser.gamal2310@mhiet.edu.eg)[1], (dr.mostafa.abd.k@mhiet.edu.eg)[2].

Abstract. In recent years, lightweight concrete has become very popular in the construction industry. And, its usage can lead to reduced costs and time. Therefore, different types of concrete (lightweight (LC), heavyweight (HC), and ordinary concrete (OC)) have been investigated to comprehend their response to earthquake load. Also, the comparison between their responses has been considered for getting the best response for various materials in construction work. The results prove that building models with LC are more resistant to earthquakes than the other types due to the reduction in the weight of buildings. Consequently, it has a role in resisting seismic forces. Furthermore, the combination of different concrete in the structural elements is another important criterion to be considered in thesis. The findings demonstrate that the responses in models for OCC-LCR (ordinary concrete for the column and light concrete for the roof) and LCC-OCR (light concrete column and ordinary concrete roof) have fewer values than the other types.

Keywords: Lightweight concrete; Heavyweight concrete; Ordinary concrete; Spectrum analysis

1. Introduction

The lightweight aggregate (LWA) has been used from an older era until now to produce the LC. And, the Romans erected the Pantheon from the pumice, which can be considered as LWA. In recent decades, they have been widely utilized throughout the world, including in the UK, Sweden, and the USA. Furthermore, the LWA has been engaged in the construction industry since 1928 until now. And, the LWA has also constructed many multistory buildings, mega-constructions, and offshore platforms [1].

The LC is utilized in constructing many structures and bridges in the USA. Also, the American Bridge, the Missouri Building, Kingston Bridge, Dubai International Airport, and the Dubai Cooper River Bridge have been built from the LC.

The concrete structures are vulnerable to earthquakes because of their mass. Therefore, the main purpose of lightweight concrete (LWC) is to reduce the dead load of a concrete structure, which enables structural designers to scale back the size of structural parts, including beams, columns, and footings.

Also, the designers have the objective of reducing the cost of construction by utilizing the LW materials. There is also potential to achieve this by decreasing the weight of the materials utilized in structural elements.

On the other hand, the LC can contribute to lowering the mass of the floor slab. Thus, the cost of construction via the LC is less than that of OC and HC.

A long time ago, there were a lot of attempts to construct buildings with different sorts of concrete. And, the LC has been employed in the concrete industry instead of OC to get significant advantages in seismic performance.

Despite there being a bunch of research for studying different kinds of LC with a variety of characteristics [[2], [3], [4],[5],[6]], there is a shortage in the investigation of their effects on seismic behavior. And few studies have focused on the behavior of the LC under seismic action. Therefore, it would be particularly interesting to examine the seismic performance of the LC. Moreover, extensive research is essentially needed to study the seismic performance of the LC building and make a comparison to other types (OC and HC). In addition, the combination between them (LC, HC, and OC) in the structural elements is another important factor to be recognized in research.

The models for twenty stories are created by ETABS to identify the seismic reaction. And, the measurements of the response that have been analyzed are story displacement, story drift, story shear force, and overturning moment.

2. Lightweight aggregate concrete properties

The LC has advantages in increasing mixture volume and reducing both density and dead weight. However, the lessening of density can contribute to enhancing thermal insulation. In addition, there are multitudes of methodologies for producing the LC, where it can be fabricated from pumice aggregate or man-made sintered aggregate (Fly ash)[1]. Also, the technology of production can be primed either by injecting air into its components or omitting the finer sizes of the aggregate, or even replacing them with a hollow, cellular, or porous aggregate. ACI Committee 213 [7] has defined the components of the LC. Where, it can be manufactured from LWA or mixed between normal fine-aggregates and lightweight fine-aggregates. Moreover, the LC is categorized by its density, which ranges from 320 to 1920 kg/m^3 via ACI Committee 213 [7].

Buildings can be constructed in low-seismic zones using lightweight concrete because of its low density. It is important to choose materials to keep the structure's dead weight low since structural seismic reactions depend on the mass or dead weight of the structure.

Furthermore, the main aim of utilizing the LWA in the concrete industry is to get an adequate strength with low density. As well as, it can avoid the shrinkage crack by maintaining the concrete element with low drying shrinkage. In addition, the numerous applications of the LC were made in the construction of long-span bridges and offshore platforms.

The LC can be sorted into three main items via the strength range [low-density concrete with a strength of 0.7-2.0 Mpa, moderate- concrete with a strength of 7–14 Mpa, and structural concrete with a strength of 17-63 Mpa][8]. Besides, this concrete has densities ranging from 300 - 800 kg/m^3, 800-1350 kg/m^3, and 1350–1920 kg/m^3, respectively[9]. Also, BS EN 206-1 [10] has determined that an oven-dry density of the Lightweight is not fewer than 800 kg/m^3 and not over 2000 kg/m^3.

What's more, the LC is sorted into fourteen classes, ranging from LC8/9 to LC80/88. In addition, the LC has a homogeneity characteristic in the components, which can be resulted from the similarity of the elastic modulus of LWA and the concrete matrix [11].

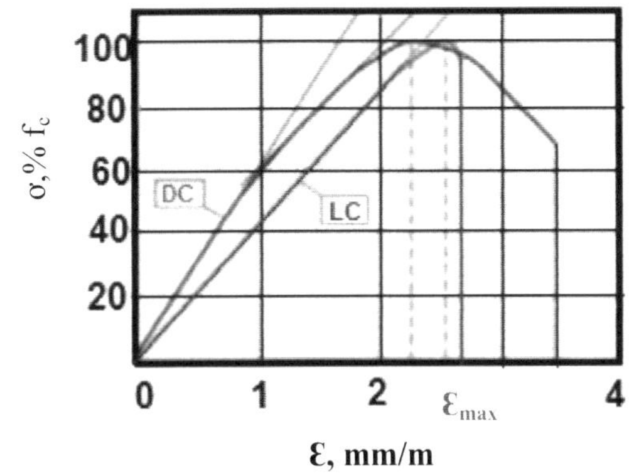

Fig. 1 Stress-strain relationship in compression test for dense concrete (DC) and LC [11]

In LWA concrete, there are three stages for the development of cracks. Where, the first crack appears when tensile stress reaches 85-90% of concrete strength. As a result, the LC can be employed in elements that have uncracked conditions under a large tension load. Fig.1 indicates the comparison between the destruction mechanism of the LC and normal concrete. Where, the stress-strain relationship has generated a rectilinear in the uniaxial compression test [11].

The crack and destruction mechanism of the LC has a different path and location than the OC. Typically, the destruction of dense concrete happens in the linkage region between cement and aggregate. So, it is considered the weak point among them. Furthermore, the linkage region can be mostly loaded due to the stress concentration resulting from significant differences in the elastic modulus of the matrix and aggregates. In addition, the destruction points of the LC occur significantly in the cement matrix owing to the high strength in the connection zone. Furthermore, the lack of stiffness and the spread of cracks forming are the biggest disadvantages of the LC[11].

The shrinkage of the LC rises by 50% over its value in OC. In addition, the rise in shrinkage can result from lowering the modulus of elasticity. Several researchers have examined increasing the shrinkage of the LC [[12],[13],[14]]. Time dependence, the outdoor temperature, and humidity conditions are different parameters that can contribute to reducing the shrinkage rates. Furthermore, the shrinkage rate can be identified in all types of aggregate. As well, the aggregate sintered at a high temperature has less value in the shrinkage than other types. Otherwise, the rate of creep in the LC is higher than in OC. Also, the creep for the higher classes of the LC is like the DC [[13],[14]]. Whereas, the impacts of lowering the density of LC can lead to poor mechanical properties compared to the characteristics of DC.

However, HC spreads widely in the structural systems. It is also crucial in the construction of critical buildings. As well as, HC has different physical and mechanical properties. And, it is categorized into fifteen diverse types depending on the different rates of barytes in concrete mixtures. In addition, the values of specific gravity for HC range from 2600 kg/m^3 to 3000 kg/m^3. Therefore, the reduction of dead load and increasing the strength capacity can be major factors in studying the LC.

The models for twenty stories are created by ETABS to identify the seismic reaction. And, the measurements of the response that have been analyzed are story displacement, story drift, story shear force, and overturning moment.

3. Research Significance

This study focuses on the examination of the seismic response demands for the LC by evaluating variety kinds of concrete (LC, HC, and OC). In addition, the combination between them (LC, HC, and OC) in the structural elements is another important factor to be recognized in research. The results include story displacement, story drift, story shear force, and overturning moment.

The models that have been used in the studies cases are introduced in Table 1, where: Model 1: LC, Model 2: OC, Model 3: HC.

Table 1 The properties of the LC, HC, and OC

Ser	Type of RC	Self-weight kg/m^3	Modulus of elasticity kg/cm^2	Concrete compressive strength kg/cm^2
LC [15]	LC	1700	$1.2*10^6$	250
HC [16]	HC	3203	$4.3x10^6$	260
OC	OC	2200	$2.1x10^6$	250

4. Modal response spectrum (RS) method

For the last century, the RS is the main idea in earthquake engineering to give proper standard methods for calculating the peak response of all feasible linear SDOF systems. In addition, the RS method provides structural dynamics knowledge to the design. Furthermore, The RS concept is incorporated into all of the design codes.

Zone 5-B is chosen as the seismic region, and its characteristics are adopted via the Egyptian seismic regions. Moreover, the type of spectrum shape is type 2.

Besides, seismic region factor = 0.3 g, and importance factor = 1. Also, the type of soil class is 'C', which denotes dense soil. The value of R expresses that the resistance of the overall base shear is completely done by the shear wall structure, and, its value is equal to five.

Based on the ECP (2008), total seismic mass is composed of "Dead Load (DL) + 0.25 LL + FL (flooring load)"[17]. Fig.2 depicts the RS curve for the studied cases.

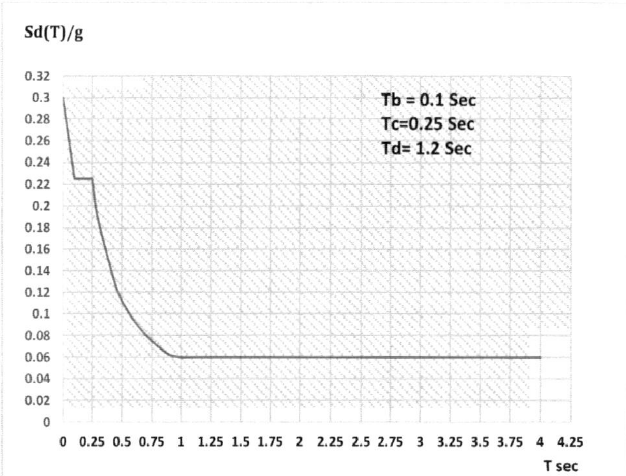

Fig.2 RS Curve

5. Modeling of Tall Building

The present study comprises twenty stories, with a total height of sixty meters. And, the typical floor height is equal to 3 m. The structural system comprises a flat slab, columns, and a shear wall system. For the analysis of modeling, it is assumed all section sizes of column and core are constant regardless of increasing the model height.

Although the highest buildings are erected on basement stories, the representative modeling of the building avoided the basement floor and considered it an ordinary floor. Moreover, the stories of the building are identical to the models' heights. The plot area of the structural building is 31.85 x 35.10 m, as shown in Fig.3.

IOP Conf. Series: Materials Science and Engineering 1269 (2022) 012004 doi:10.1088/1757-899X/1269/1/012004

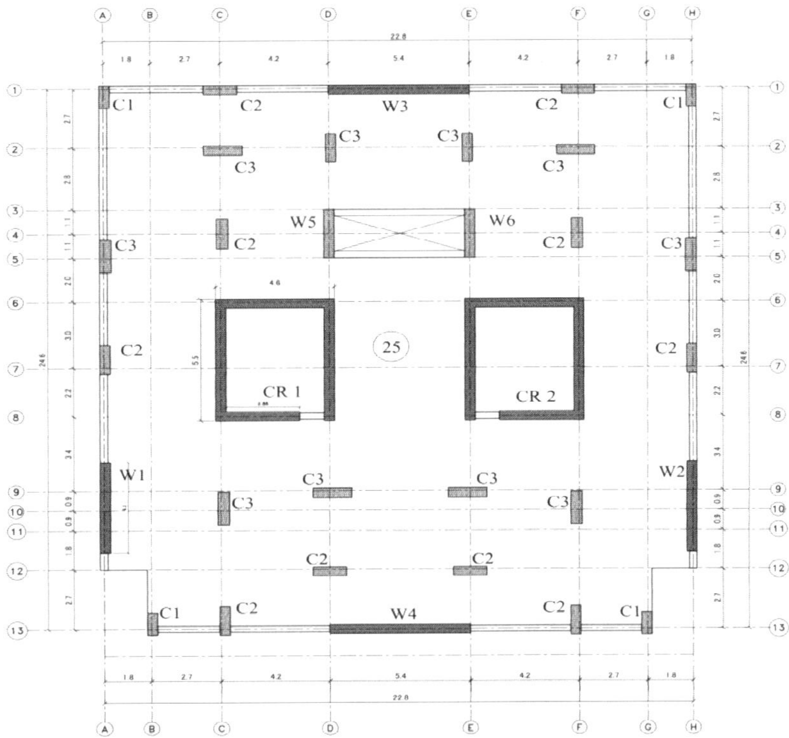

Fig.3 The configuration Plan

In ECP 2007 code [18], the conditions and the recommendations should be recognized, such as the slab thickness and the column slenderness ratio. And, various formulas are used to compute the thickness of slabs. Besides, these formulas rely on the type of construction components and the guidelines of the ECP Code. Furthermore, the deflection of the solid slabs can control the depth of the solid slabs rather than the flexural strength requirements. Therefore, it is recognized for design structural elements. All recommendation of ECP code has been take into account in the design of structural modeling.

Table 2 shows the values of Ln/t for member spans with lengths of less than 10 m. Where the Ln is a clear span and the t [the thickness of the slab] is not less than 120 mm [17].

Table 2 the values of Ln/t for member spans with lengths of less than 10 m [17]

Element	Simply support	One End continuous	Two End continuous	Cantilever
Solid Slabs	25	30	36	10
Hollow Block	20	25	28	8
Paneled Beam	T=short length Ls/(12-16)			
Flat slab	Ts is the bigger of $=\begin{cases} 150\ mm \\ \frac{L}{32} without\ drop \\ \frac{L}{36} with\ drop \end{cases}$			

The cross-section elements are chosen to be sufficient with accuracy to get a reasonable response to the structure. Moreover, a preliminary design process has been utilized to calculate the sizes of cross-section elements. Additionally, all elements are built to be safe in compliance with ECP 2007 code, and all slabs have the same thickness of 25 cm [17].

The dimensions of the exterior columns are C1 (40cm x 100cm), C2 (40cm x 130cm), and C3 (45cm x 150cm). Moreover, C2 (40cm x 130cm) and C3 (45cm x 150cm) are the dimensions of the interior columns.

As indicated in Table 3, the thickness of the walls between axes 1-4, 5-10, and 11-20 is forty centimeters. These sections have been examined for compliance with Egyptian code requirements while taking the effect of earthquake loads into account using the Egyptian code for loads and forces[17].

What's more, the dead loads comprise the self-weight and the flooring load, which is equal to 1.5 kn/m^2. The live load is equal to 2.5 kn/m^2 for each floor. The type of steel is ST40/60 with the modulus of elasticity $E = 2.0\times10^8$ kn/m^2. And, the Poisson ratio $\mu = 0.3$.

All dimensions of the cross-sections are exposed in Fig.3.

Table.3 The dimension of the cross-sections of construction elements (Beams, Columns, Shear walls)

Member	L(Length cm)	B (width cm)
	The Exterior columns (cm)	
C1	100	40
C2	130	40
C3	150	45
	The interior columns (cm)	
C2	130	40
C3	150	45
	Wall thickness (cm)	
Wall Thickness	40	

The structural model comprises a flat slab with a thickness of 25 cm. Also, the cross-section dimension of the edge beam is equal to (30 x 80) cm. Furthermore, the vertical elements of the building have composed of columns and shear walls.

The characteristics of concrete differ from one case to another according to the concrete type for each element (slabs, columns, and beams). Table 2 indicates the characteristics of concrete for the LC, HC, and OC.

6. Finite element model

A three-dimensional mathematical model of the structure has been used to calculate the critical components of the building's dynamic response. This model will accurately depict the spatial distribution of the structure's mass and stiffness. As a 3D frame structure, the building is represented by frame elements for the columns, longitudinal beams, transverse beams, and shell elements for the slabs.

7. Numerical results and discussion

- **Deformation response demands**

The values of lateral displacement are measured depending on the properties of the structural building during and after the earthquake. Therefore, it is considered a vital criterion for construction design. Otherwise, the large values of deflection and story drift can lead to a harmful effect on both of the structural or nonstructural elements [19]. So, lateral displacement is one parameter that are recognized in the study. What's more, different concrete types (LC, OC, HC) have been investigated to study their seismic performance.

Fig.4 shows the distribution of the maximum story displacement over the model's height under RES (x-Direction). Model HC displays the highest top displacement response compared to other types. And, their value is equal to 0.236 m, which is over 37.4% of OC. Besides, the maximum story displacement of the LC model is less than the OC by 18.64%.

IOP Conf. Series: Materials Science and Engineering 1269 (2022) 012004 doi:10.1088/1757-899X/1269/1/012004

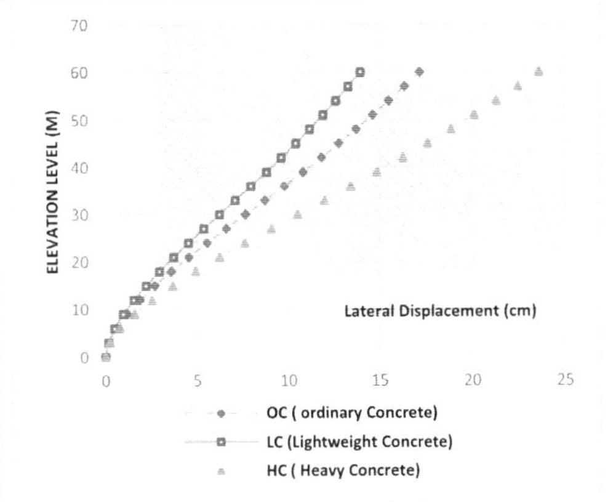

Fig. 4 Story displacement, (cm)

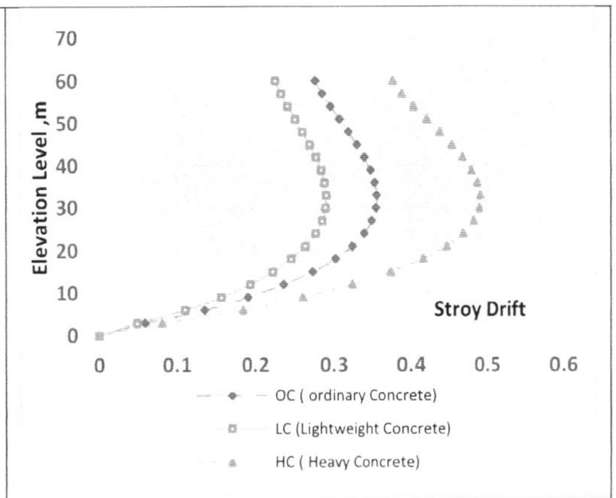

Fig. 5 Story Drift (Unitless)

The story drift response demand is investigated for previous models. And, it is compared to the reference OC (RF). The story drift over the modeling's height for diverse cases is indicated in Fig.5. The story drift ratios for models (OC-HC-LC) increase gradually over the height of the building and reach their maximum value at the 11th story level. Hence, it decreases at higher levels. In addition, the story drift ratio of the HC model is greater than the OC model, while the maximum value of the story drift for model HC is over 37.33% of the RF model. Otherwise, the decreasing percentage for LC is 18.62% compared to the RF model.

- **Base shear force response demands**

The successful comparison among OC (RF) and other types of models (LC, HC) provides a realistic simulation of shear force demand for models. Besides, the weight of construction elements for structural models impacts the shear response. So, it would be of special interest to get the best modeling in seismic resistance. Fig.6 presents the story shear response for each model.

The shear force of the LC is lesser than the RF by 17.29%. Whereas, the HC model has the biggest value and reaches more than 34.58% compared to the RF. Furthermore, the additional shear force demand of HC can violate the safe design.

- **Overturning moment response demands**

Fig.7 shows the overturning moment for each model (LC, OC, HC). The overturning moment for the HC model is more than 37.37% compared to the RF. Otherwise, the decreasing percentage in the LC model can reach 18.63% compared to that of the RF.

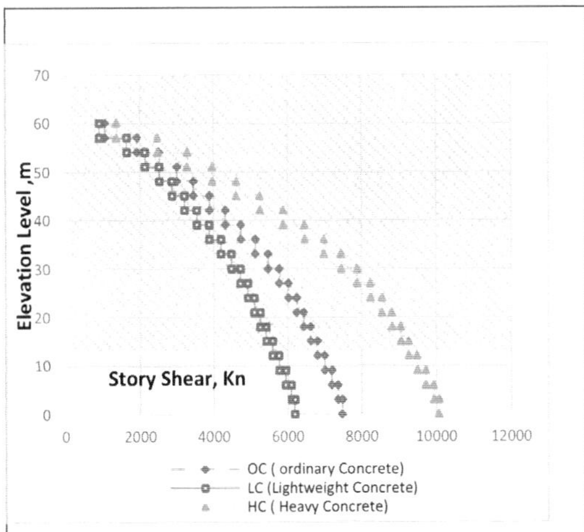

Fig.6 Story shear, (kn)

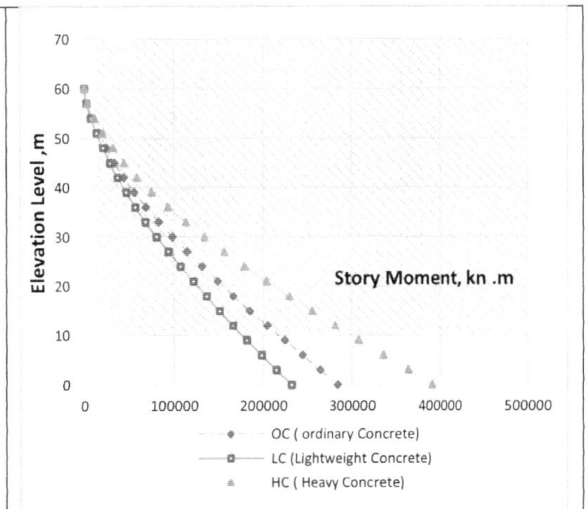

Fig.7 Story moment, (kn.m)

8. Varity of the concrete type in the structure elements

The possibility of using different concrete types (LC, OC, HC) in the structural elements is investigated in research to determine the different methodologies to improve the building's seismic performance. Fig.8 and Table.4 show different concrete types (LC, OC, HC) in the structural elements. In conclusion, the main findings are summarized as follows:

First: the lateral displacement: Fig.9 shows the distribution of maximum story displacement along model heights for studied cases. Model OCC-HCR displays the highest top displacement response compared to other types. And, their value is equal to 0.2079 m, which is over 21.03% of OCC-OCR. Otherwise, the top displacements for OCC-LCR and LCC-OCR are fewer than 8-10% of that OCC-OCR.

Second: the story drift: The story drift over the building's height for different models has been presented in Fig.10. The value of the maximum story drift of model OCC-HCR is greater than that of RF (OCC-OCR) by 20.95%. Whereas, the maximum story drift values of models OCC-LCR and LCC-OCR are fewer than 8-10% of that OCC-OCR model.

Third: the story shear: The value of the maximum base shear of the OCC-HCR model is greater than the RF by 19.09%. While, Both the OCC-LCR and LCC-OCR models have values less than those for OCC-OCR by percentages ranging from 7.5-9.5%. Fig.11 indicates the base shear force for each model under seismic loading.

Fourth: the overturning moment: Fig. 12 shows the overturning moment response for each model. The value of the maximum overturning moment of the OCC-HCR is greater than the OCC-OCR by 20.96%. Whereas, the decreasing percentage in both the OCC-LCR and LCC-OCR models ranges from 8 to 10% compared to that of OCC-OCR.

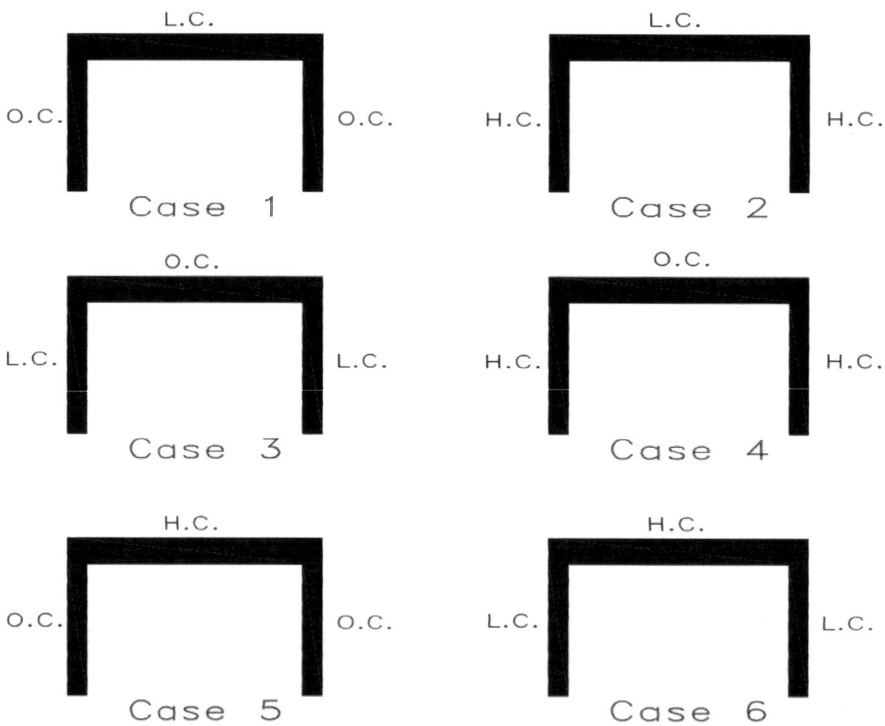

Fig.8 The cases of models

IOP Conf. Series: Materials Science and Engineering 1269 (2022) 012004 doi:10.1088/1757-899X/1269/1/012004

Table 4 the cases of studied

Ser	Case	Description	Model Code
1	Case No 1	OC column LC Roof	OCC-LCR
2	Case No 2	HC column LC Roof	HCC-LCR
3	Case No 3	LC column OC Roof	LCC-OCR
4	Case No 4	HC column OC Roof	HCC-OCR
5	Case No 5	OC column HC Roof	OCC-HCR
6	Case No 6	LC column HC Roof	LCC-HCR
7	RM (Reference Model)	OC column OC Roof	OCC-OCR

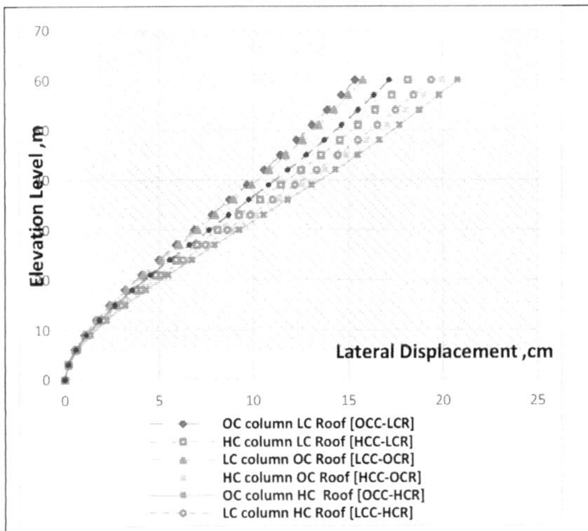

Fig.9 Lateral displacement, (cm)

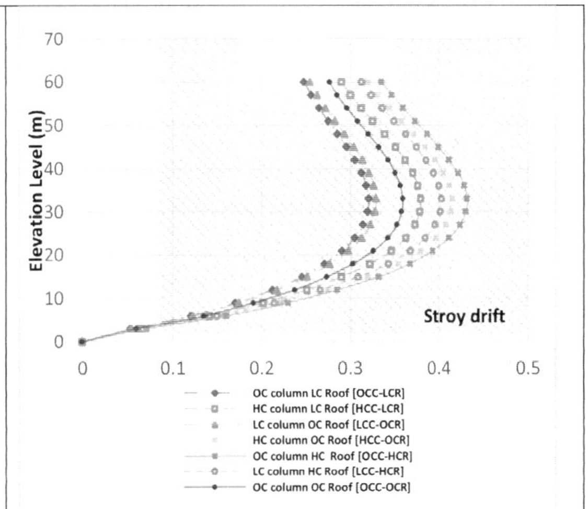

Fig.10 Story drift

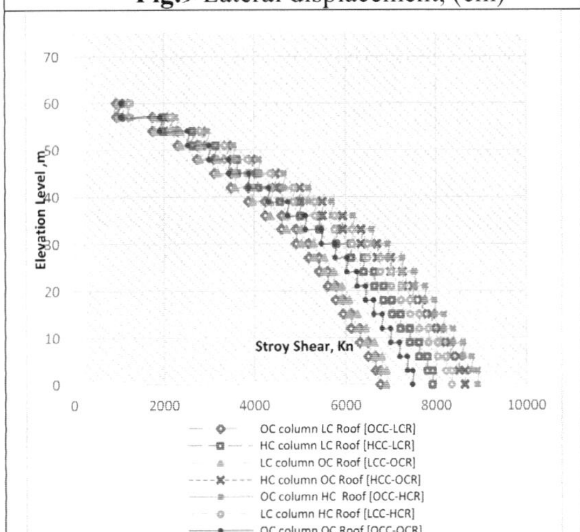

Fig.11 Story shear, (kn)

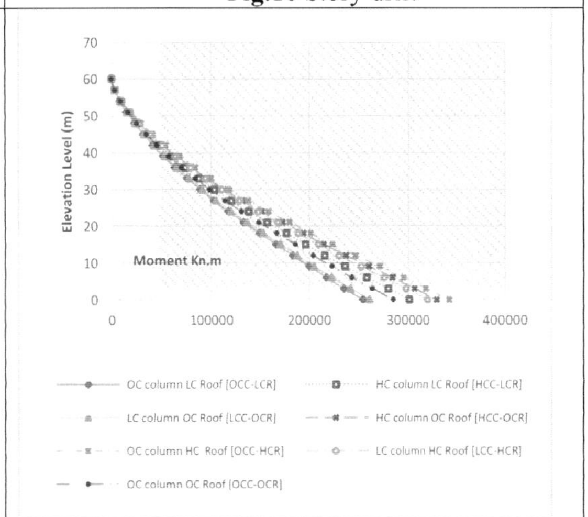

Fig.12 Overturning moment

9. Conclusion

Functionally, the structural seismic analysis depends on the dead load and the earthquake forces that are operating on it.

The usage of diverse types of concrete has a substantial effect on the seismic response for various reasons, such as the rigidity and weight of the building. The controlling of structural performance can be realized by different sorts of concrete (LC, HC, OC). So, the choice of concrete type should be considered not only for the safe design but also for the economic factor.

The findings indicate that using LC can cause a better seismic response than other types of concrete. Also, this behavior is caused by the weight of the LC, which is lower than other types [HC, OC]. Hence, it can impact on the earthquake forces. Besides, the design of the structure for story drift and story displacement of the LC model is safer than other types due to the lowering of Young's modulus of LC.

Another important factor is studying the combination of different kinds of concrete in the structure modeling (OCC-LCR (case 1) - HCC-LCR (case 2)- LCC-OCR (case 3) - HCC-OCR (case 4) - OCC-HCR (case 5)- LCC-HCR (case 6)–OCC-OCR (Reference Model)) in the elements. Nevertheless, the construction of the building with the same concrete type generate up the best behavior but also using different concrete types (LC, OC, HC) in the structural elements. In conclusion, the findings illustrate that the lowest response can be gotten in both OCC-LCR (case 1) and LCC-OCR (case 3) models compared to other types, as clarified in Fig. 9, Fig. 10, Fig. 11, and Fig. 12.

Less bending moments and shear forces were recorded for the LWC structure during the seismic investigation, which may allow for a reduction in the cross-section of the members. Practically, the explanation for these changes is a decrease in Young's modulus of lightweight concrete.

On the whole, the reduction of the column rigidity generates up the appropriate behavior for the energy dissipation of the seismic response, hence improving seismic performance.

REFERENCES

[1] Mindess S, Young JF and Darwin D 2003 *Concrete, Prentice Hall, Upper Saddle River.*

[2] Kilic A, Atis CD, Yasar E and Ozcan F 2003High-strength lightweight concrete made with scoria aggregate containing mineral admixtures. *Cement and concrete research* **33** No.10 p 1595.

[3] American Concrete I, Committee. 1977 Recommended practice for selecting proportions for structural lightweight concrete : (ACI 211.2-69).

[4] Kan A and Demirboga R 2009 A novel material for lightweight concrete production. *Cement and Concrete Composites* **31** No.7 pp 489-95.

[5] Vandanapu SN and Krishnamurthy M 2018 Seismic performance of lightweight concrete structures *Advances in Civil Engineering* **2018**.

[6] Wang S, Wu X, Yang J, Zhao J and Kong F 2020 Mechanical behavior of lightweight concrete structures subjected to 3D coupled static–dynamic loads *Acta Mechanica* **231** No.11 pp. 4497-511.

[7] Committee ACI, American Concrete I. 2014 Guide for structural lightweight-aggregate concrete.

[8] Mindess S. Developments in the Formulation and Reinforcement of Concrete 2019 *Woodhead Publishing.*

[9] TotalBoox, Tbx. 2014 Eco-efficient Masonry Bricks and Blocks.

[10]Institution BS. BS EN 197-1: 2011 Cement. Composition, specifications and conformity criteria for common cements. (London: British Standards Institution).

[11]Szydlowki R and Mieszcak M 2017 Study of Application of Lightweight Aggregate Concrete to Construct Post-tensioned Long-span Slabs *Procedia Engineering Procedia Engineering* **172** pp. 1077-85.

[12]Kahn LF, Lopez M. 2005 Prestress Losses in High Performance Lightweight Concrete Pretensioned Bridge Girders *pcij PCI Journal* **50** No.5 pp. 84-94.

[13]Domagala L. 2020 Durability of Structural Lightweight Concrete with Sintered Fly Ash

Aggregate. Materials Materials **13** No.20 pp. 45-65.
[14]Tian Y, Shi S, Jia K and Hu S. 2015 Mechanical and dynamic properties of high strength concrete modified with lightweight aggregates presaturated polymer emulsion. *JCBM Construction and Building Materials* **93** pp. 1151-6.
[15]Deifalla A. 2020 Design of lightweight concrete slabs under two-way shear without shear reinforcements: A comparative study and a new formula. *JEST Engineering Structures* **222**.
[16]Topçu IB. 2003 Properties of heavyweight concrete produced with barite. Cement and Concrete Research Cement and Concrete Research **33** No.6 pp. 815-22.
[17](ECP) HaBNRC. 2008 Egyptian code of practice for loads on buildings and bridges-ECP-201. Cario: In: Ministry of housing. Utilities and Urban Communities.
[18](ECP) HaBNRC. 2007 Egyptian code for design and construction of reinforced concrete structures-ECPCS-203. Cairo: In: Ministry of housing. Utilities and Urban Planning.
[19]Abdel Raheem S E, Ahmed M M and Alazrak T M A 2015 Evaluation of soil-foundation-structure interaction effects on seismic response demands of multi-story MRF buildings on raft foundations. *Int J Adv Struct Eng International Journal of Advanced Structural Engineering (IJASE)* **7** No.1 pp. 11-30.

Experimental Study for the Comparison between Core, Schmidt and Cubic Concrete Compressive Strength After Durability Test

Mostafa Abd Elrazek [1], M M Rashwan [2] and Mohammed H M Ahmed[3]
[1] Lectures of properties and strength of materials, Civil Eng. Department, Higher Institute of Engineering and Technology, New Minia, Egypt
[2] Prof. of properties and strength of materials, Civil Eng. Dept., Faculty of Eng., Assuit University.
[3]Demonstrator at Higher Institute of Engineering and Technology, New Minia, Egypt

E-mail: (dr.mostafa.abd.k@mhiet.edu.eg) [1], (mohamed.rashwan1@eng.au.edu.eg) [2], (pro_mohamed@mhiet.edu.eg)[3].

Abstract. Estimating the compressive strength of concrete exposed to different environmental conditions is an essential parameter in judging the quality of concrete and proving its ability to perform as planned. There are various destructive (DT) and non-destructive (NDT) methods for estimating compressive strength. Some NDT and DT procedures, such as Schmidt hammer and core testing, have been used for estimating concrete strength for a long time. These methods consider indirect and predicted tests to evaluate concrete strength in existing structures. Various factors influence these tests, so it is difficult to precisely assess the hardened concrete strength. Thus, this research aims to identify a coherent relationship that connects the results of these tests and correlates them with the results of cube compressive strength by assessing data from laboratory tests conducted under control and durability conditions using statistical techniques. To achieve this purpose, three grades of concrete were considered C_{250}, C_{450}, and C_{600} Kg/cm^2. Six concrete cubes of size (15x15x15 cm) and four concrete columns of size (30x15x50 cm) were cast for each grade and tested to assess the cube, Schmidt hammer, and core compressive strength of concrete. Half of the columns and cubes were cured in drinking water (control condition) and the other half of the samples were subjected to durability cycles for 92 days (durability condition). In the durability condition, the samples were exposed to 10% sodium sulfate solution for a day, placed in a water tank for a day, heated to 100°C at a laboratory furnace for a day and then left at average laboratory temperature for a day to finish one cycle. The results were tabulated and graphed to notice any differences between DT, NDT and cube compressive strength tests. The results showed that the cube strength is higher than the core strength for all grades of concrete under control and durability conditions. The cube test indicates a lower compressive strength than the Schmidt hammer test method at grade C_{250} under control condition and greater compressive strength in the rest of the grades.

1. Introduction

For engineers and academics, determining the concrete's compressive strength in existing structures is a major challenge. Concrete structures are regularly subjected to many factors that affect their strength, like natural environmental conditions and industrial chemicals [1]. These variables can cause long-term damage to concrete, which decreases its compressive strength and shortens its service life or durability and life expectancy. Over the years, engineers have been concerned about how to evaluate the strength of finished structures and predict their behavior in various environmental conditions.

There are many test methods to predict the compressive strength of concrete. Codes and standards still favor cube and cylinder compressive strength testing [2]. Cubes are manufactured alongside concrete elements; these test samples were cast and tested in a controlled environment until the testing day. However, in outdoor environments, concrete strength can deteriorate over time, resulting in decreased reliability and overall structural performance [3]. So, in most circumstances, estimation of the concrete's in situ compressive strength at a specific location in the structure is required [4].

Various tests are available to determine existing concrete elements, ranging from entirely NDT tests that cause no damage to the concrete to DT techniques that require the surface to be repaired after the test [5]. These tests aim to manage productivity and identify under-service structural on-time deterioration.

The Schmidt hammer test is the most commonly used NDT among those available. Because of its simplicity, portability, low cost, and non-destructiveness, as well as its speed and ease of implementation. It has been utilized all over the world to assess concrete strength [6].

A surface's hardness can be assessed using the Schmidt hammer. It relies on the concept that an elastic mass's ability to rebound is directly related to how hard the concrete surface it strikes is. A spring-controlled mass with consistent energy is made to hit the concrete surface when the plunger of a rebound hammer is pressed against it and then rebounds back. The hardness of the concrete surface affects how much the mass will rebound [7].

The DT most frequently used to assess the in-situ compressive strength of concrete is the coring test [8]. While the rebound hammer test provides a more empirical indication of the quality of the concrete, the core sample and testing provide a more direct evaluation of strength. Rotary cutting tools with diamond bits are regularly utilized to cut concrete cores. These cores were sometimes extracted with embedded reinforcing elements. The cores were then capped with molten Sulphur to make the ends plane and parallel before being tested [9].

The relationship between cube and core compressive strength was discussed by Dr. Suresh G. Patil et al. [10] With a w/c of 0.55 and M-200 grade concrete, twelve concrete cubes (15x15x15 cm) were cast. Five slabs with dimensions of 50 cm x 50 cm and a thickness of 12 cm were created using the same w/c ratio. Using cores with different H/D ratios (1.5, 1.75, and 2.0) and diameters of 50 mm, the relation between the cube strength and the strength determined using those cores was investigated in an experimental investigation. A total of thirty cores were taken out of the slabs; fifteen of them were taken out of the concrete part that contained no reinforcement, and the other fifteen were taken out from the concrete part that contained reinforcement. The compressive strength estimated from concrete cubes reveals stronger strength than the estimated core strength for all H/D ratios, according to the study data. Comparing the core's compressive strength to that of cube strength, the cores with reinforcement with H/D ratio of 2.0 demonstrated a higher level of strength. The development of correlation equations indicated that for cores with H/D ratios of 1.5 and 1.75, the coefficient of correlation between the core strength V.S. slabs rebound number (R=0.955) and between the estimated strength of cube from core strength V.S. slabs rebound number (R=0.929) is nearly the same, but for cores with H/D ratio 2.0, the coefficient is lower (R=0.850). The correlation coefficient between the actual cube strength V.S the strength of the core and between the actual cube strength V.S the estimated strength of the cube from core strength was almost the same for cores with H/D ratios of 1.5 and 1.75 (R=0.939 and R=0.942, respectively), while it was lower for H/D ratio 2.0 (R=0.843).

The relation between the core strength and the strength of the cube was also described by M. Yaqub et al. in their paper [11]. Sixteen cores (diameter 7.5 cm) and twelve cores (4.8 cm diameter) were taken from a 12-year-old building's hardened concrete. Hardened concrete cubes of two sizes (15x15x15 cm) and (10x10x10 cm) were created from stone pieces removed from structural parts of the structure using a stone cutting machine. The strength of cores and cubes was evaluated using a compression testing machine. The results showed that the compressive strength of cores (75 mm diameter) was 0.693 times that of cubes (15x15x15 cm). Cores (7.5 cm diameter) have a compressive strength that is 0.528 times that of cubes (10x10x10 cm). The core's compressive strength (4.8 cm diameter) is 0.724 times that of cubes (15x15x15 cm). Cores (4.8cm diameter) have a compressive strength that is 0.552 times that of cubes (10x10x10 cm).

In order to achieve reliable data for the Schmidt hammer test method, Mostafa Abd Elrazek and Yasser Abdel Shafy Gamal [12] aimed to show how to combine NDT with machine compression testing (MCT). In order to investigate the NDT and compare it to the Machine Compression Test (MCT), various concrete compositions with different proportion mixes of cement, water content, and aggregate were used in all tests, which were created for different grades of reinforced concrete ranging from medium to high, from Grade 200 to 800 Kg/cm^2. All findings were collated and graphically depicted to detect any differences between the DT and MCT. Using regression analysis, they establish a linear mathematical relation between concrete compressive strength and rebound number. Regression analysis was used to analyze the variance between tests. In the case of medium and semi-high grades, the null assumption was valid according to statistical analysis; nevertheless, there were variations in very high grades.

Adil M. EL-Bashir et al.[13] investigated the application of DT and NDT procedures in concrete, as well as compared DT and NDT testing. Two grades of concrete were considered: 250 and 300 Kg/cm^2. Three beams and twenty-four cubes were cast and tested for each grade. Two DTs (crushing and tensile strength) and two NDTs were conducted (Schmidt hammer test and ultra-sonic test). After the casting, these tests were conducted at various intervals (7, 14,28 and 56 days). Compression, Schmidt hammer and ultrasonic tests were carried out on the cubes. Tensile strength was tested on the beams. According to the designated compressive strength, the best results were obtained after a 28-day crushing test. According to known concrete graphs, DT provides more logic and better graph forms than NDT.

DT and NDTs were used by Jedidi Malek et al. [14] to study the mechanical properties of concrete. With variable cement dose and w/c ratios, eight different concrete proportions were chosen to produce cylindrical specimens (16x32 cm). Then, concrete is tested at various ages (7, 14 and 28 days) for compressive strength. The results showed that the compression resistances obtained by the compression test were higher than those obtained by the rebound hammer test at the ages of 7 and 14 days. At the ages of 7 days and 14 days, the average differences between the two techniques were 14% and 17%, respectively. There was a reversal of the situation at 28 days; compression test resistances were lower than rebound hammer test resistances, with an average difference of 5%. The concrete's maturation could account for this result. The compressive strength and homogeneity of the concrete can be evaluated in situ using the Schmidt hammer test without using a core, which is more applicable to older concretes than to more recent concretes.

In order to assess the compressive strength of recycled aggregate concrete (RAC), Mostafa Kazemi et al. [15] used Schmidt hammer and core tests (RAC). This study's objective was to apply these techniques to calculate the compressive strength of RAC and then contrast the findings to those obtained from 15 cm cube specimens. For the Schmidt hammer, core and cube tests ninety-six cube specimens and eight concrete slabs were prepared. Additionally, correlations between core and Schmidt hammer tests were achieved by combining different methods. A two-variable equation with a high correlation coefficient (R^2 = 0.99) was created using the specimen results: $f_c = 173\,C + .229\,R - 2.755$ where (f_c)is the predicted compressive strength, (C) is the Core strength with L/D ratio = 2, and (R) is the rebound number. The results demonstrated that, for estimating the compressive strength of RAC and standard concrete, a multi-variable equation based on the outcomes of these tests is more promising than a single-variable equation based on the Schmidt hammer results.

Nilam Bhosale and P.A. Salunkhe [16] studied the relationships between the Schmidt hammer, ultrasonic, core, and cylinder compressive strength of concrete. An experiment was conducted using DT and NDT methods on various concrete mixtures, including M200, M250, and M300. For each concrete grade, a slab measuring (200x100x20 cm) was cast, and cores were taken from the slab. To compare the findings, (10x20 cm) cylinders were also cast. Compressive strength was measured on cubes of (15x15x15 cm). The results led to the conclusion of some relationships, such as the relationship between the compressive strength of cylinders and cores, which was given by the formula $f_{cor} = -.034{f_{cyl}}^2 + 2.586f_{cyl} - 19.25$, the relationship between the cylinders' and cores' Schmidt strengths, which was given by the formulas $R_{core} = -.020{R_{cyl}}^2 + 2.152R_{cyl} - 16.75$ and $R_{core} = 27.51 \ln R_{cyl} - 64.57$

The relationship between Schmidt strength and cores strength which was $R_{core} = -.050{f_{cor}}^2 + 3.487f_{cor} - 31.16$ as well the relationship between Schmidt strength and cylinder strength which was $R_{cyl} = -.037{f_{cyl}}^2 + 2.712f_{cyl} - 19.85$ also concluded in this study.

Venkata Rambabu V and Amit B Mahindrakar [17] studied the behavior of concrete and its ability to resist several environmental conditions, these conditions cause concrete to deteriorate by causing shrinkage cracks, reduction of strength at later ages, a decrease in strength, high porosity, corrosion of steel, and a reduction in life span or durability. To address these issues, a variety of concrete additives, such as Fly-ash, Silica fume and other additives, as well as Pozzolana cement, slag cement and low calcium fly ash-based concrete, were used. Based on this study weathering agents will influence the properties of freshly poured and hardened concrete, necessitating more analysis to ensure that concrete can perform its intended function. According to this research, ordinary Portland cement can be replaced with Portland pozzolana cement, although this requires further investigation. Epoxy resins could be utilized to overcome corrosion issues, but more research is needed. To meet the requirement, admixtures must be added. Concrete with fewer pores will be better for the environment. The long-term durability of the concrete will be significantly impacted by the type of curing utilized in marine locations.

Despite the many methods and techniques, there is a deficiency in estimating the compressive strength of existing structures exposed to different environmental conditions, which were taken care of by this study and for which the study was conducted. In-situ compressive strength is still one of the less well-understood concrete characteristics that are used on a regular basis, so this study examined the relationships between the compressive strengths of concrete cubes, cores, and Schmidt hammer in different environmental conditions. The results of compressive concrete tests will be easier to analyze and interpret as a result of this work.

2. Experimental

2.1. Materials

The properties of the materials used in the laboratory to make concrete mixtures were as follows:

2.1.1. Cement. The cement used was Al Tameer Ordinary Portland Cement (OPC), with a CEM I (42.5 N). Cement's physical properties are shown in Table 1.

Table 1. Physical properties of Cement

Properties	Results	Limitation of ESS*
Specific Gravity	3.15	————
The fineness of cement using Blaine apparatus (cm^2/gm)	3505.42	Not less than 2750 cm^2/gm
Setting time (min)		
Initial sitting time	110	Not less than 45 min.
Final sitting time	316	Not more than 10 hr.
Compressive strength of cement mortar after 7 days (Kg/cm^2)	347	Not less than 270 kg/cm^2
Water required for cement paste of standard consistency (cm)	111 (w/c = .2775)	————
Le chatelier expansion of cement (mm)	1.4	————

ESS*: Egyptian Standard Specification

2.1.2. Fine Aggregate. Coarse sand was used. The physical properties of fine aggregates, as well as the results of sieve analysis, are shown below in table 2 and table 3.

Table 2. Results of Fine Aggregate's Sieve Analysis

Sieve size (mm)	Cumulative % pass	Limitation of ESS* %
10	100	100
5	97.4	89-100
2.5	93.55	60-100
1.25	87.25	30-100
.63	65.7	15-100
.31	11.2	5-70
.16	1.5	0-15

ESS*: Egyptian Standard Specification

Table 3. Physical Properties of Fine Aggregate

Properties	Results
Specific gravity	2.5
Volume weight (t/m^3)	1.62
Amount of clay and fine dust (%)	2.36
Fineness modulus	2.44
Void ratio (%)	35
Specific surface area (cm^2/gm)	52.24

2.1.3. Coarse Aggregate. Basalt aggregate was used as coarse aggregate in concrete mixes. Tables 4 and 5 show the sieve analysis results and the used basalt's physical characteristics.

Table 4. Results of coarse Aggregate's Sieve Analysis

Sieve size (mm)	Cumulative % pass
10	100
5	91.975
2.5	30.5
1.25	7.67
.63	1.23
.31	.625
.16	.625

Table 5. Physical Properties of Coarse Aggregate

Properties	Results
Specific gravity	2.7
Volume weight (t/m^3)	1.57
Fineness modulus	4.67
Basalt crushing value	14
Void ratio (%)	42
Specific surface area (cm^2/gm)	6.74

2.1.4. Additives

- *Fly ash*

Fly ash is a modern concrete additive that comes in spherical and fine powder forms. It lowers the amount of water needed for concrete and has a lubricating effect that makes concrete flow and pumps more easily. The concrete is also more cohesive and less likely to segregate. The specific gravity of fly ash is 2.2 gm/cm^3 and its optimum dosage is 25-35% by weight of cement.

- *Sikament -NN*

A powerful liquid superplasticizer with a dual action that can either create free-flowing concrete or operate as a significant water-reducing agent to encourage high early and ultimate strengths. The density of the product is 1.2 gm/cm^3. Depending on the intended workability and strength, the product's dosage ranges from 0.6-3% by weight of cement.

- *Sikament -R 2008*

a superplasticizer that is highly effective and has a setting retarding effect for making free-flowing concrete in hot temperatures. Additionally, a significant water-reducer supports high early and ultimate strengths. The density of the product is 1.2 gm/cm^3. Depending on the intended workability and strength, the product's dosage ranges from 0.6-3% by weight of cement.

2.2. Mix design

Concrete mix design of grades C_{250}, C_{450}, and C_{600} kg/cm^2 were prepared. The mix proportions of concrete per cubic meter are listed in table 6.

Table 6. Concrete Mix Proportions

Concrete grade (kg/cm^2)	C_{250}	C_{450}	C_{600}
Cement (kg/m^3)	350	450	500
Water (kg/m^3)	167.7	156.82	150
Basalt (kg/m^3)	1201.5	1128.7	1032
Sand (kg/m^3)	801	752.5	688
Fly ash (kg/m^3)	-	22.5	25
Sikament -NN (kg/m^3)	-	6.7	7.5
Sikament -R 2008 (kg/m^3)	5.25	9	10

2.3. Testing and Methods

2.3.1. Cube compressive strength test. For each grade, six concrete cubes (15x15x15 cm) were cast. Three of the cubes were cured in drinking water for 28 days, and the other three samples were exposed to durability cycles to study the behavior of concrete under harsh environmental conditions. The concrete samples in the durability cycle were exposed to 10% sodium sulfate solution for a day, placed in a water tank for a day, heated to 100°C at a laboratory furnace for a day and then left at average laboratory temperature for a day before the cycle was repeated for 28 days.

2.3.2. Destructive and non-destructive tests. Four concrete columns (30x15x50 cm) without reinforcement were cast for each grade to perform Schmidt hammer and core testing (Three cores were extracted from each column). Half of the columns were cured in drinking water for 92 days, while the other samples were put through 92 days of durability cycles (one day in 10% sodium sulfate solution, one day in drinking water, one day in a 100° C furnace, and one day in average laboratory temperature).

IOP Conf. Series: Materials Science and Engineering 1269 (2022) 012005 doi:10.1088/1757-899X/1269/1/012005

3. Results

3.1. Cube compressive strength test

Characteristic compressive strength is measured using cube specimens. The cubes are tested in a compression testing machine with a capacity of 150000 kg. The load that causes the cube specimen to fail is observed. The average compressive strength of concrete cubes is presented in table 7 below.

Table 7. Results of cube compression test

Grades	**Curing**	**Average compressive strength (Kg/cm^2)**
C_{250}	Cured in drinking water	279
	Exposed to durability cycles	173
C_{450}	Cured in drinking water	466
	Exposed to durability cycles	283
C_{600}	Cured in drinking water	629
	Exposed to durability cycles	367

3.2. Schmidt hammer test

This technique is based on the idea that the hardness of the surface that an elastic mass impacts determine how much it will rebound. The hammer's plunger is forced firmly and steadily against the concrete surface at right angles until the spring-loaded mass is released from its locked position. The mass's distance traveled is determined as a rebound number, which is expressed as a percentage. A low rebound value suggests that the concrete is fragile and has a soft surface. High rebound values show that the concrete is tough and durable.

For each grade, the rebound numbers were taken randomly from four columns, two columns of them cured in drinking water and the other two columns exposed to durability cycles. They were subjected to an initial load of about 1/3 of the anticipated ultimate load to simulate the pre-loading of concrete in existing structures. With the Schmidt hammer in the horizontal position, totaling 20 readings are taken at different places in each column element (10 readings on each side). Table (8) shows each grade's average rebound number and cube compressive strength.

Table 8: Results of Schmidt hammer test

Grade	Curing	Average Rebound number	Average compressive strength (Kg/cm^2)
C_{250}	Cured in drinking water	35.5	322.5
	Exposed to durability cycles	16.35	-
C_{450}	Cured in drinking water	35.77	329
	Exposed to durability cycles	16.3	-
C_{600}	Cured in drinking water	40.5	411
	Exposed to durability cycles	23.2	138

IOP Conf. Series: Materials Science and Engineering 1269 (2022) 012005 doi:10.1088/1757-899X/1269/1/012005

3.3. Core testing

Cores were extracted from the direction of a 150 mm thick cast column after the Schmidt hammer test. As a result, in this case, it was decided to use cores with a diameter of 75 mm. The ratio of height to diameter (L/d) was 2. three Cores were extracted from each of the four columns for each grade. A total of twenty-four cores were extracted from the concrete columns that had been marked.

Using a diamond saw with water lubricant, the extracted cores were cut into parallel, straight edges and then capped with Sulphur to make the ends smooth. Cores were tested in the concrete laboratory's compression testing machine after 24 hours of capping. The average core results for each grade are shown in table 9.

Table 9. Results of Core test

Grade	Curing	Average core compressive strength (Kg/cm^2)
C_{250}	Cured in drinking water	248
	Exposed to durability cycles	102
C_{450}	Cured in drinking water	325.5
	Exposed to durability cycles	196
C_{600}	Cured in drinking water	443
	Exposed to durability cycles	285

4. Discussion

From observation of tables, a comparison of average cubic, Schmidt hammer and core compressive strength for C_{250}, C_{450} and C_{600} grades of concrete was done.

4.1. Control Condition

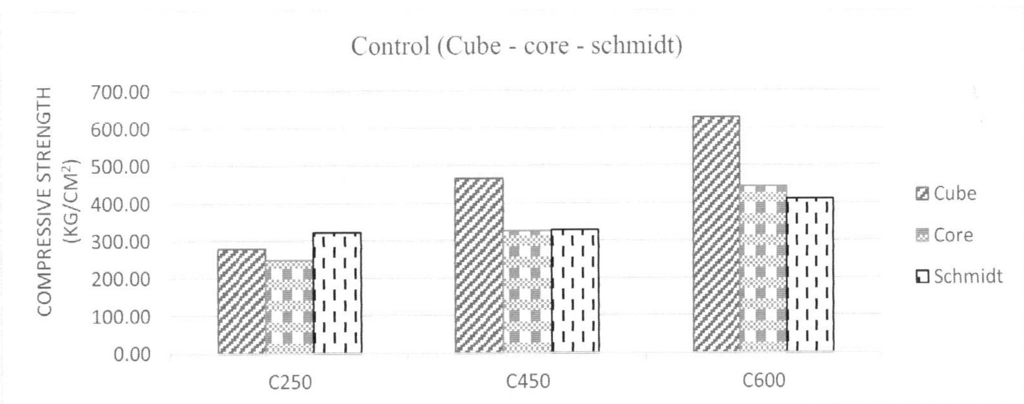

Figure 1. The average compressive strength of concrete under control condition as evaluated by destructive (Core), non-destructive (Schmidt hammer) and cube compressive strength tests.

In Figure 1, the outcomes of the cube test are compared with those of the Schmidt hammer and core tests and the differences between them are expressed as percentages. At grade C_{250}, it was found that the resistances obtained by the cube compressive test are lower than those obtained by the Schmidt hammer, with an average difference of 15.6%, while in the other grades, the cube compressive strength shows higher results than the Schmidt compressive strength which is 41.7% and 53% at grade C_{450} and C_{600} respectively.

IOP Conf. Series: Materials Science and Engineering 1269 (2022) 012005 doi:10.1088/1757-899X/1269/1/012005

It is also observed that the resistances obtained by cube compressive strength were higher than those obtained by core compressive strength in all three grades. The compressive strength of cubes is 12.4%, 42% and 42% higher than the compressive strength of cores at grades C_{250}, C_{450} and C_{600}, respectively. As we can see, the difference in compressive strength between cube and Schmidt hammer test methods is minimal at grade C_{250} and the difference increases as the grade increases. It might be explained that the Schmidt hammer test results are correct up to a point. So, they are not always taken as the final word on any structural decisions.

It is also observed that the compressive strength obtained by the cube test is higher than the compressive strength obtained by the core test in all three grades of concrete.

4.2. Durability Condition

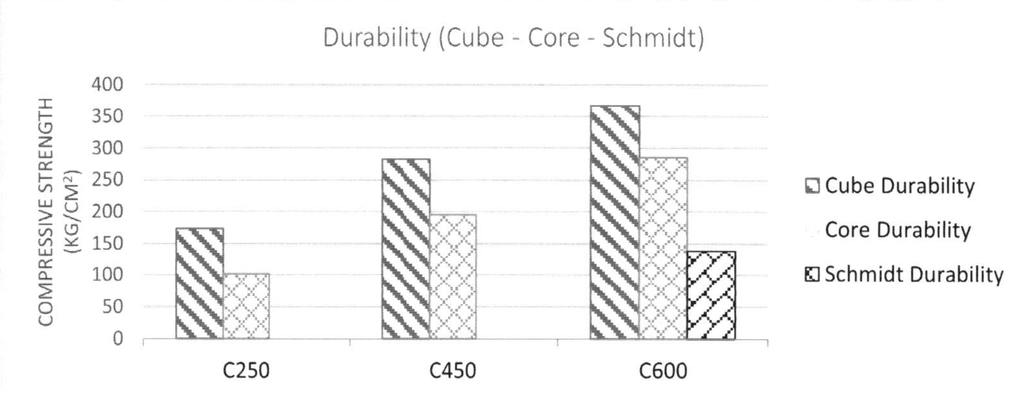

Figure 2. The compressive strength of concrete under durability condition as evaluated by destructive (Core), non-destructive (Schmidt hammer) and cube compressive strength tests.

In figure 2, The results show that the Schmidt compressive strength test at grades (C_{250} & C_{450}) failed because the average rebound number is less than 20, indicating that the concrete is influenced by the durability cycles in these grades affecting the quality of concrete surface hardness. It is also observed that the cube strength was higher than the core and Schmidt hammer strength of all three concrete grades. The cube compressive strength is 69.6%, 44.5% and 28.7% higher than the Core compressive strength. At grades C_{250}, C_{450} and C_{600}, respectively. On the other hand, cube compressive strength is 166% higher than Schmidt compressive strength.

4.3. Comparison between control and durability condition

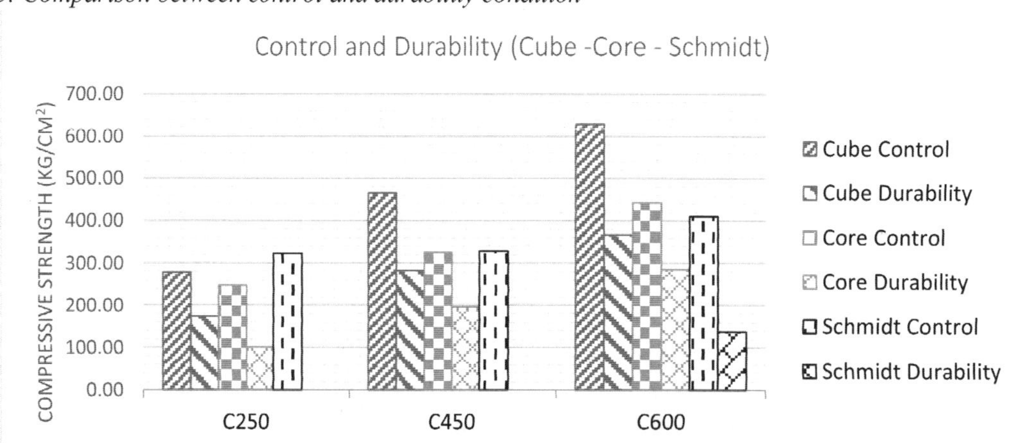

Figure 3. The compressive strength as evaluated by destructive (Core), non-destructive (Schmidt hammer) and cube tests for three grades of concrete under control and durability conditions.

By analyzing the results from figure 3 to find out the influence of durability on Cube, Core and Schmidt compressive strength test, we can see that; at grade $C_{250,}$ the average cube strength at the control condition is about 61% higher than the durability condition and the average core strength at the control condition is about 143% higher than the durability condition.

At grade $C_{450,}$ the average cube strength at the control condition is about 64.6% higher than the durability condition and the average core strength at the control condition is about 66.3% higher than the durability condition.

At grade $C_{600,}$ the average cube strength at the control condition is about 71.4% higher than the durability condition, the average core strength at the control condition is about 55.3% higher than the durability condition and the average Schmidt strength at the control condition is about 197.7% higher than the durability condition.

It can also be observed that at grade $C_{250,}$ the cube compressive strength under the control condition is 173% higher than the core compressive strength under the durability condition. At grade $C_{450,}$ the cube compressive strength under the control condition is 137.9% higher than the core compressive strength under the durability condition. At grade $C_{600,}$ the cube compressive strength under the control condition is 120.5% higher than the core compressive strength under the durability condition.

Also, the results shown in fig show that the Schmidt hammer test at grades (C_{250} & C_{450}) gave a low rebound number (less than 20) which demonstrates the concrete's weakening and soft surface because the quality of concrete was affected by the durability cycles. At grade $C_{600,}$ the average cube strength is 355.7% higher than the average Schmidt strength.

4.4. Statistical Analysis of the Experimental Result using OriginPro 2018

This section's effort was put forth to establish equations that fit the relationship between cube, Schmidt hammer, and core compressive strength. These equations allow predicting concrete's standard cube compressive strength using DT and NDTs. Statistical regression analysis was performed using OriginPro2018 software to explain the results of the tests and to obtain empirical equations that can predict the cube concrete strength from core compressive strength and Schmidt compressive strength under control and durability condition, as shown in Figures (4&5&6).

Despite the limited number of specimens, the entire range of anticipated compressive strengths was covered. For each of the three grades of concrete, a correlation between the average results of three cubes, six cores, and 20 Schmidt hammer readings was undertaken to develop a coherent formula representing the relationships between the tests. However, it is clear that more samples would greatly improve the amount of data the models could provide.

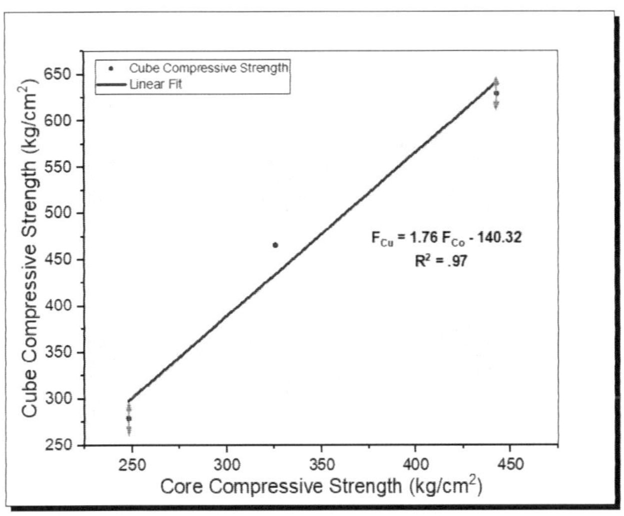

Figure 4. Correlation between Cube and Core compressive strength under control condition

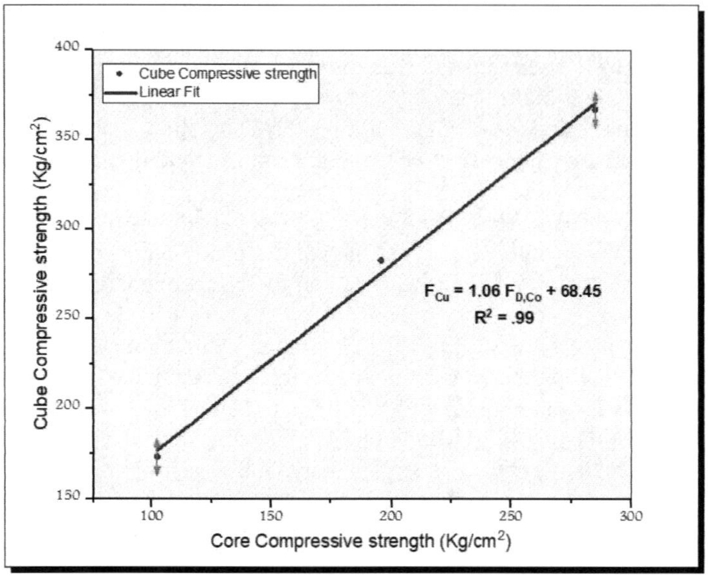

Figure 5. Correlation between Cube compressive strength under control condition and Core compressive strength under durability condition

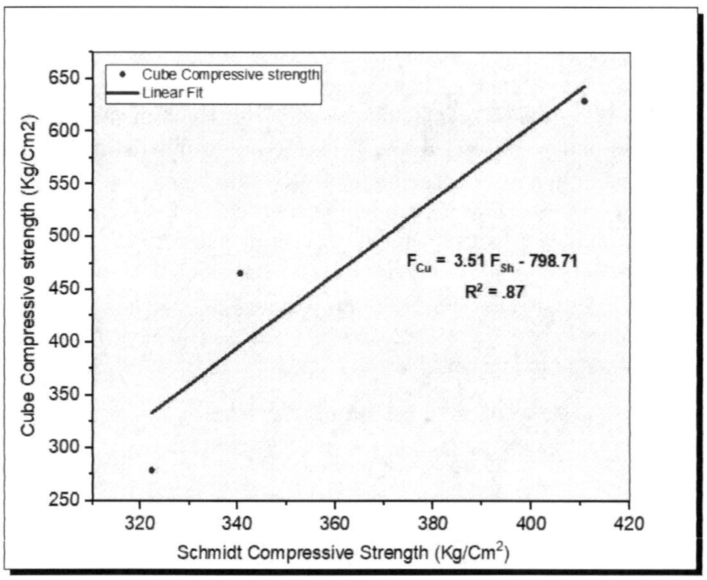

Figure 6. Correlation between Cube and Schmidt compressive strength under control condition

The results were randomly selected and used to obtain the equations:

(i) The relation between cube and core compressive strength under control condition is:

$$f_{cu} = 1.765 f_{co} - 140.32 \quad (kg/cm^2)$$

where f_{cu} is the predicted cube strength under control condition and f_{co} is the core strength with L/d ratio = 2 under control condition. R-square (R^2) obtained for this equation is 0.95, meaning that we can explain 95% of the variability in the data around the regression line. And 5% of the residual data could not be explained by this equation. This indicates that the correlation curve and the variable data are closely related. Figure 4 shows the equation.

(ii) The relation between cube compressive strength under control condition and core compressive strength under durability condition is:

$$f_{cu} = 1.059 f_{D,co} + 68.457 \quad (kg/cm^2)$$

where f_{cu} is the predicted cube strength under control condition and $f_{D,co}$ is the core strength with L/d ratio = 2 under durability condition. R-square obtained for this equation is 0.99. Figure 5 shows the equation.

(iii) The relation between cube and Schmidt compressive under control condition is:

$$f_{cu} = 3.511 f_{sh} - 798.715 \quad (kg/cm^2)$$

where f_{cu} is the predicted cube strength and f_{sh} is the Schmidt strength. R-square obtained for this equation is 0.87. Figure 6 shows the equation.

The obtained results demonstrate that the proposed equations can be applied to estimating cube strength on concrete types like those covered above, considering the possibility that even small variations in the composition of the concretes could result in significant errors in DT and NDT strength estimations. This means that developing your own regression model is always appropriate. Additionally, it should be emphasized that the established formulas might not be applicable to concretes with various compositions or concrete exposed to different circumstances.

5. Conclusions

The compressive strength of concrete at different grades under control and durability conditions was assessed using the Schmidt hammer, core, and cube testing. The experimental study led to the following conclusions:

- The Cube compressive strength of concrete specimens under control curing conditions shows higher results than Schmidt hammer compressive strength at higher grades (C_{450} & C_{600}) by 41.7% and 53%, respectively. While in the lower grade (C_{250}), the cube compressive strength shows lower results than Schmidt compressive strength by 15.6%. It is also worth noting that as the concrete grade increases, the difference between the cube and Schmidt compressive strength under the control condition increases from 15.6% at grade (C_{250}) to 53% at grade (C_{600}). It is possible to explain that the Schmidt hammer test results are accurate up to a certain point. So, they are sometimes not always taken as a final call for any structure-related decisions.

- The use of the rebound hammer test method on concrete exposed to aggressive environmental conditions is unsuitable for determining its strength. The durability condition affected the quality of the concrete surface, which led to the failure of the Schmidt hammer test at lower grades (C_{250} & C_{450}) and also caused a big difference in compressive strength between the Cube and Schmidt hammer at a higher grade (C_{600}).

- When compared to the measured core compressive strength for all grades, the compressive strength obtained from the cubes indicates greater strength at control and durability conditions, ranging from 12.4% to 69.6%. It is also observed that the difference in compressive strength between cube and core test methods under durability condition is greater than the difference under control condition at grades C_{250} and C_{450}, while at C_{600}, there has been a reversal of the situation. It is also worth noting that as the concrete grade increases, the difference between the cube and core compressive strength under control condition increases from 12.4% at grade (C_{250}) to 42% at grade (C_{600}), while at durability condition, the difference decreases from 69.6% at (C_{250}) to 28.7% at (C_{600}).

- When compared between normal and high strength concrete, Schmidt hammer and Core test methods are more effective tools for determination of strength in normal strength concrete class.
- The difference between the strength of cubes under the control condition and the strength of cubes under the durability condition increases as the concrete grade increases from 61% at grade (C_{250}) to 71.4% at grade (C_{600}). In contrast, in the core test, the difference decreases as the concrete grade increases from 143% at (C_{250}) to 55.3% at (C_{600}).
- Empirical equations that can predict the compressive strength of cube from core under control and durability conditions and Schmidt hammer under control condition derived from a statistical analysis of the data gathered from the experimental tests.

$$f_{cu} = 1.765 f_{co} - 140.32$$ ***(kg/cm²)*** core under control condition

$$f_{cu} = 1.059 f_{D,co} + 68.457$$ ***(kg/cm²)*** core under durability condition

$$f_{cu} = 3.511 f_{sh} - 798.715$$ ***(kg/cm²)*** Schmidt under control condition

- There is a strong correlation between Cube strength, Schmidt hammer and Core obtained from concrete columns, which is presented by the good coefficient of determination R-square which ranges from (.87 to .99).
- The study also demonstrates that the strength of concrete is influenced by the environment to which it is subjected.

6. References

[1] Salhi M, LI A, Ghrici M, Bliard C and Boubekeur T 2019 Effect of Acids and Sulfates on The Durability of Self-Compacting Concretes *1st Int. Conf. on Materials, Environment, Mechanical and Industrial Systems ICMEMIS'19* (Id: hal-02367771).

[2] Atoyebi1 O D, Ayanrinde O P and Oluwafemi J 2019 Reliability Comparison of Schmidt Rebound Hammer as a Non-Destructive Test with Compressive Strength Tests for Different Concrete Mix *J Phys.: Conf. Ser* **1378** 032096.

[3] Weilai Y, Shiyong J, Wei F & Tao C 2017 Correlation between the Compressive, Tensile Strength of Old Concrete under Marine Environment and Prediction of Long-Term Strength *Hindawi Advances in Materials Science and Engineering* **2017** (Id: 8251842).

[4] Ryan D and Rankin G I B 2012 Cube, Cylinder, Core and Pull-off Strength Relationships ICE *Institution of Civil Engineers Structures and Buildings* **166** pp 521-536.

[5] Şebnem K, Aydın B and Ercan I 2020 The Relationship Between Concrete Strengths Obtained by Destructive and Non-destructive Methods *Iranian Journal of Science and Technology, Transactions of Civil Engineering* **44** pp 91–105.

[6] Ferhat A and Mehmet S 2010 Correlation between Schmidt Hammer and Destructive Compressions Testing for Concretes in Existing Buildings *Scientific Research and Essays* **5** (13) pp 1644–1648.

[7] Aydin A and Basu A 2005 The Schmidt Hammer in Rock Material Characterization *Engineering Geology* **81** pp 1–14.

[8] Shafik K, Ali A A and Ahmed G 2014 Reliability of Core Test – Critical Assessment and Proposed New Approach *Alexandria Engineering Journal* **53** pp 169–184.

[9] Karthikeyan V, Loganathan S and Venugopal P 2013 A Study on RC Columns and Slabs and Restoration of RC Columns of an Existing Multistoried Building *International Journal of Engineering Innovation & Research* **2** Issue 2 (ISSN: 2277–5668).

[10] Suresh G P and Shivakumar 2017 Correlation between Actual Compressive Strength of Concrete and Strength Estimated from Core *IOSR Journal of Mechanical and Civil Engineering* **14** Issue 2 ver III pp 27-44.

IOP Conf. Series: Materials Science and Engineering 1269 (2022) 012005 doi:10.1088/1757-899X/1269/1/012005

[11] Yaqub M, Anjum M J 2006 Comparison of Core and Cube Compressive Strength of Hardened Concrete *31st Conference on Our World in Concrete and Structures* (Id: 100031053).
[12] Mostafa A and Yasser A G 2021 The Reliable Concrete Compression Strength Assessment by Schmidt Hammer for Different Concrete Grades *IOP Conf. Ser.: Mater. Sci. Eng.* **1171** 012004.
[13] Adil M and Salih M A 2015 Destructive and Non-Destructive Testing for Concrete in Sudan-A Comparative Study *IOSR Journal of Mechanical and Civil Engineering* **12** Issue 6 ver VI pp 27–44.
[14] Jedidi M and Machta K 2014 Destructive and Non-destructive Testing of Concrete Structures *Jordan Journal of Civil Engineering* **8** No 4 pp 432–441.
[15] Mostafa K, Rahmat M and Jorge de Brito 2019 Compressive strength assessment of recycled aggregate concrete using Schmidt rebound hammer and core testing *Construction and Building Materials* **224** pp 630–638.
[16] Nilam B and Salunkhe P A 2016 To Establish Relation Between Destructive and Non-Destructive Tests on concrete *Inter. Journal of Engineering Research and General Science* **4** Issue 2 pp 634–644.
[17] Venkata R V and Amit B M 2017 Impact of Aggressive Environment on Concrete – A review *Inter. Journal of Civil Engineering and Technology* **8** Issue 9 pp 777–788.

Investigation of Solar Energy Potential and PV-Outputs in Rural and Desert Areas: Case Study Egypt

Ashraf S. Khamees[a], U. Ali Rahoma[a], A. H. Hassan[a], Tarek Sayad[b] and Mostafa Morsy[b]

[a] *Solar and Space Research Department, National Research Institute of astronomy and geophysics, 11421 Helwan, Cairo, Egypt.*

[b] *Astronomy and Meteorology Department, Faculty of Science, Al-Azhar University, Cairo, 11884 Egypt.*

E-mail: eng.ashraf.khamees@gmail.com

Abstract. This study aims to identify the best strategic sites in Egypt available to build the PV systems needed to support required electricity to the country and the desert areas, in addition to deduce the total monthly, annual electricity budget and their generation costs. Egypt has a high solar energy potential which is crucial for sustainable development, as it receives 3450 hours of solar radiation (SR) throughout the year with a high intensity ranged from 8.5KWh/m^2/d in southwest part to 2.1 KWh/ m^2/d at the north coast. The observed SR at four stations (Asyut, Aswan, Kharga, Cairo) were used to assess the accuracy of the ERA5 reanalysis dataset. Moreover, seven strategic sites were chosen across the country including the Western Desert (El-Golf, Habatta, and Sewa), the Eastern Desert (Saffaga, Shiab-Elbanat), and the Sinai Peninsula (El-Agma and El-Arish) to identify the most suitable stations for the construction of PV systems. Three different types of PV cells with different efficiency (η1=19.1%, η2=22.6% and η3=22.4 %) were utilized to compute the energy that could be generated at the lowest cost at the seven selected sites across Egypt. The results showed that the monthly total cost from η1 was 0.011 $/W during June and exceeded 0.015 $/W in December, while the annual total costs are ranged from 0.0119 $/W in El-Golf and 0.01494 $/W in El-Arish. Also, the result for η2 was ranged from 0.01$/W to 0.026 over El-Golf and Shiab-Elbanat, while the last cell η3 was recorded high cost over all sites but gives more energy. Finally, it is concluded that El-Golf, Shiab-Elbanat, and Saffaga are the best three sites with the lowest coast for building a PV system followed by Habatta and El-Agma, while the energy generation costs at both El-Arish and Sewa were the highest.

Keywords: Solar panels equation of the cost - solar radiation - ERA5 Reanalysis dataset - PV generating system - electricity consumption.

1. Introduction

For solar energy applications like photovoltaic energy use, solar energy is a clean, renewable, and sustainable energy source [1, 2]. Solar energy has been mostly selected for increased development throughout the previous few decades. In fact, many individuals believe that, in the not-too-distant future, solar energy will be able to meet the majority of our energy needs [3]. A renewable resource that has become widely used in both small- and large-scale electricity productions is solar energy. Either non-concentrated photovoltaic (PV) modules or concentrated solar thermal (CST) systems can produce electricity from sun irradiation. Global irradiance on the plane of the PV modules, which may be determined from global horizontal irradiance, is used to calculate PV output [4];[5]. Shortwave light and ultraviolet energy are two types of energy emitted by the Sun [6]. Some of it is reflected back to space by clouds, some is absorbed by the atmosphere, and some is absorbed at the Earth's surface when it reaches the Earth [7]. The sun produces solar energy, which is a combination of light and heat. This energy travels from the sun to the earth, where it is collected by humans using solar collectors and converted into any desired type of energy. This renewable energy source is powerful enough, according to an estimation, to replace the electricity we get from 650 barrels of oil every year [8].

Many practical situations require the calculation of local solar irradiation [9]. Estimating local sun irradiation is critical for a variety of practical issues, including the right design of building energy cooling loads and solar energy systems. Such as building energy cooling load design, solar energy systems (thermal or photovoltaic), agriculture, and building energy cooling load irrigation Global energy demand has consistently increased in recent decades, owing to fast population expansion. In terms of energy, costs in the Region are high [10]. Market fragmentation results in greater costs and less competition [7]. The long-term data of monthly average global solar radiation (SR) are essential for designing and sizing of systems [11]. Therefore, many research papers and scientific articles focused on the presenting of SR distribution and statistics either as maps based on the measured data on earth surface or from satellite imaging [12]. In some studies, the amount of SR was quantified by developing correlations or models based on a set of sky clearness index and weather data like sunshine duration ratio, mean temperature, and relative humidity.

The process of directly converting sunlight into electricity using solar cells is known as photovoltaic. It is a quickly expanding and increasingly essential renewable alternative to conventional fossil fuel energy generation today, but it is still a relative newcomer when compared to other electricity generating technologies, with the first viable photovoltaic systems exhibited in the 1950s [13]. The space industry, which required a power supply separate from "grid" power for satellite applications in the 1960s, gave solar research and development a first major boost [14]. Solar cells were thousands of times more expensive than they are now, and the perceived need for an alternative to grid power was still a decade away, but they became an intriguing scientific variation to the rapidly expanding silicon transistor development, with several potentially specialized niche markets [15].

The copper mining business requires three types of energy: power on (electricity), heat, and fuel. Electricity has a 35 percent share, diesel has a 26 percent part, and biomass has a 20 percent stake [1]. Energy expenses are now considerable, accounting for around 25–30% of total operating costs. Energy access is a fundamental requirement for economic development and a critical element for accelerating economic growth [16]. Photovoltaic energy generation is still believed to be developing around the world, with increasing in the market, especially since the year 2000. The key driver of this expansion has been cost reduction due to technological advancements. Furthermore, the minimal environmental effect, modular nature, and low maintenance costs have all contributed to this rise, which has been mostly driven by grid-connected systems, which began rapidly rising around the year 2000 [17]. From 2000 to 2011, the amount of electrical energy generated by solar technologies increased by 20% every year [18, 19]. A reduction in the price of electricity generated is essential to sustain and even exceed this growth rate.

The weather, as well as the cycle of days and seasons, influences the production of a solar power plant. Outside of the control of plant operators, it ranges from zero to full power [20]. As repeated winter crises demonstrate, relying on too few external supply increases vulnerability. Energy security will be improved by a wider diversity of supply via interconnections and true regional markets [21]. It is critical to improve efficiency, which includes energy conservation and the use of additional renewable energy sources [22]. [23] Combat the harsh environment; most Arab countries rely on electricity mostly for air conditioning and water generation. This is accomplished by burning natural gas and light hydrocarbon fuels, both of which are non-renewable and harmful. When a location is equipped with a pyranometer that has been operational for several years, it may be easier to harness solar energy resources. If no local measurements are available, however, approximation approaches must be used to predict solar radiation characteristics [24]. As a result, the data from the climate reanalysis systems, like ECMWF-ERA5, was used in this investigation because it is excellent at representing solar radiation and atmospheric parameters [25].

In 2003, Japan produced approximately half of the world's solar cells. The United States had a share of 12%. PV cell manufacturers produced 742 Megawatts of product in 2003, according to the supply side. Japan has surpassed the US as the world's largest net exporter of PV cells and modules. Electric power systems are built to meet a large number of different consumers' aggregated instantaneous electricity demand [26].The current plan in Egypt is to install a capacity of 2.3 GW of solar powered plants by 2017 [27]. More than any other energy technology over that time period, photovoltaic (PV) costs have decreased by a factor of about 100 [28, 29 & 30]

In order to expand existing electricity production capacities and satisfy expanding demand, Egypt's demographic trends necessitate intelligent long-term planning of energy sector investments at the national level. Egypt has one of the most ideal climates for the world's largest renewable energy generation. As a result, the region's solar power has showed commercial traction in a developing clean energy export market. Thus, this study aims to identify the best strategic sites in Egypt available to build the PV systems needed to support required electricity to the country and the desert areas, in addition to deduce the total monthly and annual electricity budget and their generation costs using three PV modules.

2. Study area

Egypt encompasses urban, desert, coastal and vegetated surface conditions. Egypt has a large solar energy potential, and harnessing it is vital for national sustainable development through efficient energy planning and progressive fossil-fuel independence. Egypt has four distinct main regions, namely, the Nile Valley, the Delta, the Western Desert, the Eastern Desert, and the Sinai Peninsula [31]. The Nile Valley and Delta region is divided into two sectors: Upper Egypt and Lower Egypt, where Upper Egypt extends from Wadi Halfa in the south to Cairo in the north, while Lower Egypt extends from Cairo to the Mediterranean coast. The Western Desert covers the area west of the Nile Valley to the Libyan border in the west and from the Mediterranean coast in the north to the borders of Sudan in the south. The Eastern Desert extends to cover the area east of the Nile Valley to the Red Sea in the east and from Lake Manzala in the Mediterranean to the southern borders of Egypt and includes the Red Sea Mountain chains [32]. The Sinai Peninsula resembles a triangle; its base extends along the Mediterranean Sea in the north and its apex point is at Ras Mohammed on the Red Sea coast from the south, while it is bordered on the east by the Gulf of Aqaba and on the west by the Gulf of Suez and Suez Canal. To achieve the main purposes of this study, four stations and seven strategic stations distributed to cover the climatic regions of Egypt with varying topography and different weather conditions are selected as shown in Figure 1.

Egypt is known for its abundant natural resources (SR), ranging from 5.74 KWh/m^2/d (El-Arish) to 6.74 KWh/m^2/d (El-Golf). Even more, the maximum recorded annual mean sunshine duration ranges from 9 to 10 h in winter and 13 to 14 h in summer. When compared to European countries, these data are at least three times larger.

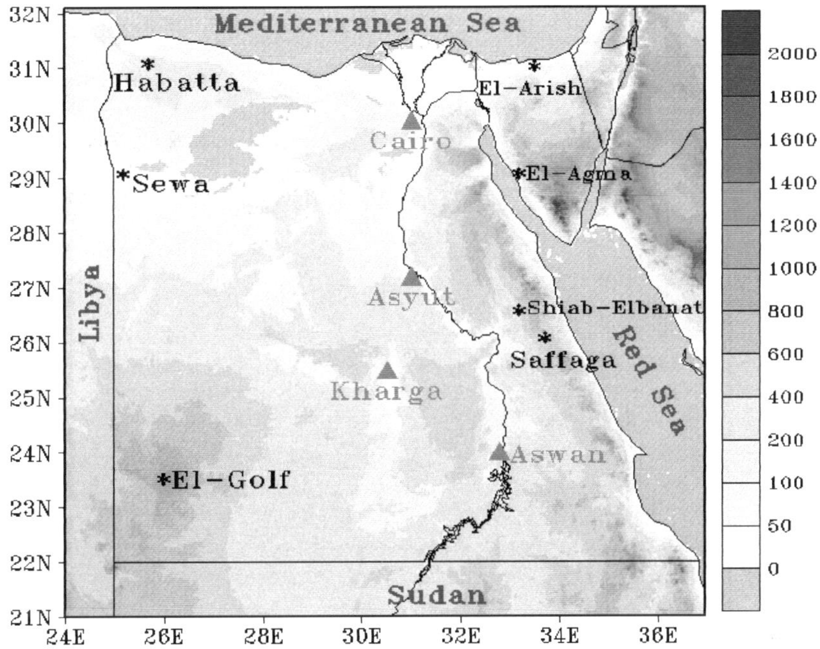

Figure 1. Topographic map of Egypt and the selected four stations (red triangle) and seven strategic sites (black stars).

The location and elevation of the selected stations and sites is demonstrated in Table 1. The seven strategic sites are chosen based on several factors, the most important of which is the geographical nature and elevations, as it is characterized by vast and uninhabited areas, which contributes effectively to the implementation of the solar cell system. El-Agma and El-Arish sites cover the Sinai Peninsula, while El-Golf, Habatta, and Sewa sites cover the Western Desert, and both Saffaga and Shiab-El-banat sites cover the Eastern Desert. Moreover, Cairo, Asyut, Kharga and Aswan stations are used to analyze and interpret the solar radiation data.

Table 1. Station of study, its location and its weather

Stations	station			station location
	Latitude (°N)	Longitude (°E)	Elevation (m)	
Asyut	27.16	31	234.7	Middle Egypt
Aswan	23.97	32.8	200	Upper Egypt
Cairo	30	31	34.4	Cairo Area
Kharga	25.45	30.5	77.8	Western Desert
El-Agma	29	33	595	Desert
El-Arish	31	33.5	10	Mediterranean
El-Golf	23.44	25.8	1260	Desert

Habatta	31	25.5	182	Desert
Sewa	29	25	200	Desert
Saffaga	26	33.7	153	Coast
Shiab-Elbanat	26.5	33	650	Coast

3. Data Used

Due to the lack of solar monitoring stations and scarcity of solar radiation data available at the selected seven strategic sites, the solar radiation data from the fifth-generation atmospheric climate reanalysis produced by European Centre for Medium-Range Weather Forecasts (ECMWF-ERA5) is used. The obtained hourly surface solar radiation downwards (SSRD) parameter from ERA5 (https://cds.climate.copernicus.eu/) is converted to daily and monthly data series. The daily observed global solar radiation data obtained from the Egyptian Meteorological Authority (EMA) at Cairo, Asyut, Kharga and Aswan stations are used to assess the accuracy SSRD from ERA5 during the period 2011-2019.

4. Methodology

The Grid Analysis and Display System (Grads) is used to display the spatial distribution of ERA5's annual SSRD and its corresponding annual energy cost across Egypt.The Data Analysis and Graphing Software (Origin) is also used to compute the summary and descriptive statistics of SSRD and display the different analysis curves. Furthermore, three common PV modules with different efficiencies (Table 2) are used to estimate the available energy and the cost of a PV system. The PV module involves several PV cells connected in parallel to increase current and in series to produce a higher voltage. The PV cell dimensions in M1 (Poly crystalline), M2 (Mono crystalline), and M3 (Mono crystalline) are 156.75 x 157.75, 158.75 x 158.75, and 156.75 x 156.75 mm respectively, while their power range is 4.47–4.67, 5.44–5.69, and 5.23–5.47 respectively. Also, the price of the PV cell in M3 (0.159$/Wp) is higher than in M2 (0.135$/Wp) and M3 (0.0946$/Wp) as indicated in Table 2. These modules are used under standard solar irradiance (SI = 1000 W/m2), solar reference spectrum (AM = 1.5), and cell temperature (T = 25 °C) conditions. Thus, the total cost of each PV module can be estimated according to Eq. 1 [33].

Table 2. Different commercial common modules under our studied tested.

Module	M1	M2	M3
Cell type	Polycrystalline	Monocrystalline	Monocrystalline
Series name	156.75-POLY-5BB-PID	JS158M5 Mono	5BB Mono PERC Cell 158.75
Cell efficiency (η)	18.3-19.1 %	21.6 – 22.6 %	21.4 – 22.4 %
Power range	4.47 – 4.67Wp	5.44 – 5.69Wp	5.23 – 5.47Wp
Cell thickness	200MM	190MM	170MM
Region	China	China	China
Dimensions	156.75*157.75MM	158.75*158.75 mm	156.75*156.75mm
Price	0.0946$/Wp	0.135$/Wp	0.159$/Wp

$$C = \frac{P}{G \times \eta / m} \quad (1)$$

Where G is the incident solar radiation (SR) at a site, m is the sunshine duration (h), C is the price of one watt equal, μ is the efficiency (%) and P is the price of the PV cell. The south facing flat plate tilted at latitude Φ is processed as in Eq. 2 [34] .

$$\mathbf{Cos}\,(\beta) = \mathbf{Cos}\,(\alpha)\mathbf{Cos}\,(\mathbf{H}) \quad (2)$$

Where β is the angle between the beam and the normal to the plate, α is the solar declination for the time, and H is the hour angle. Also, the energy from the panel can be estimated using some information about the area of the panel (A in m^2), global solar radiation (SR in $KWh/m^2/d$), performance ratio (P as loses of the cable dust losses of the battery charge) and finally the most important parameter is the efficiency (η in %) as shown in the Eq. 3. Where, the efficiency of a PV cell is determined as the fraction of incident power which is converted to electricity.

$$\textbf{Energy} = \textbf{RS} * \textbf{A} * \textbf{P} * \eta \qquad (3)$$

5. Results and Discussion

The scatter plot and the summary statistics of solar radiation data between the observed values and estimated ERA5 values at the selected four stations (Cairo, Asyut, Kharga, and Aswan) are performed to investigate the accuracy of ERA5 as demonstrated in Figure 2 and Table 3. The summary statistics include the maximum, minimum, mean, standard error, standard deviation, the correlation coefficient (R^2), and the coefficient of determination (R). The R^2 values are 0.9821, 0.949, 0.9853, and 0.9743 for Asyut, Aswan, Cairo, and Kharga, respectively, with very good correlation (R > 97 %). Furthermore, the SE and SD produce minimum values for all stations, with the SE for Asyute, Aswan, Cairo, and Kharga being 0.026, 0.021, 0.03, and 0.023, respectively, and the SD being 1.29, 1.23, 1.75, and 1.32.Accordingly, these results indicate that the estimated solar radiation from ERA5 corresponds well and with high accuracy to the observation and therefore can be used at the other seven selected strategic sites in Egypt.

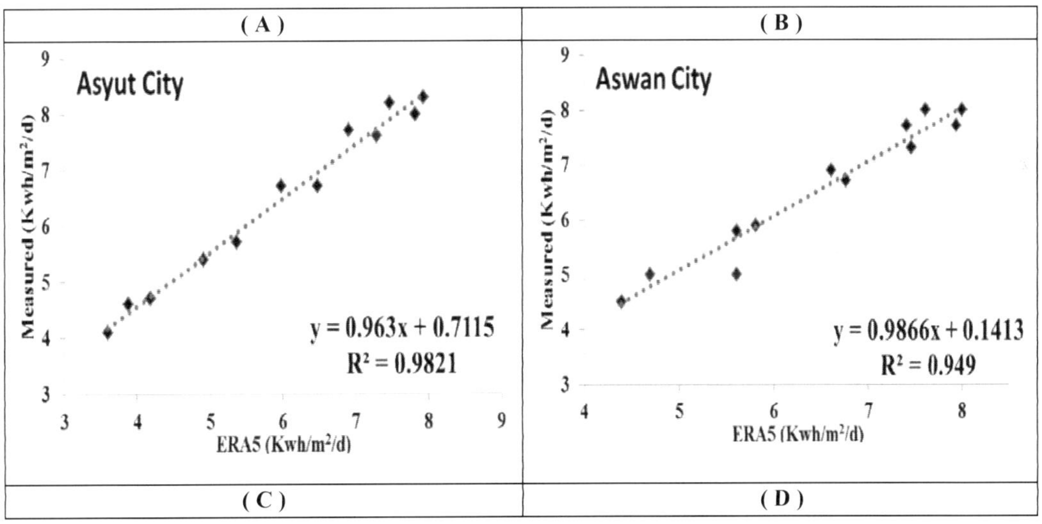

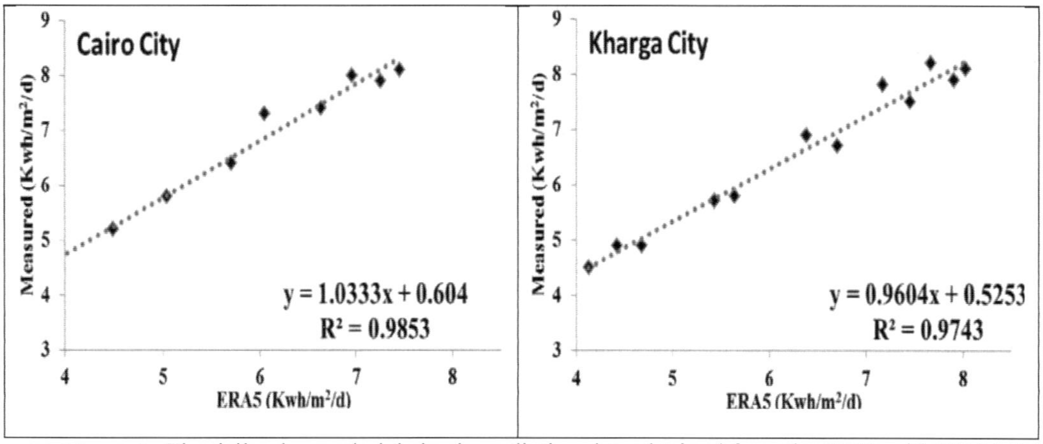

Figure 2. The daily observed global solar radiation data obtained from the EMA with ERA5 [34].

To study the variation of solar radiation over Egypt, the data from ECMWF-ERA5 was chosen over the years 2011–2019. The first class of this station is found in the Sinai Peninsula Area and the variation of SR over these two stations is shown in figure 3 In table 3, El-Agma Station has a higher SR value than El-Arish, but the two are close in the summer. The daily SR over El-Agma station ranged from 3.6 to 8 KWh/m^2/d, and the daily SR over El-Arish station ranged from 3 to 7.9 KWh/m^2/d.

Table 3. Summary statistics between observed and the estimated ERA5.

Station	Asyut		Aswan		Cairo		Kharga	
Statistics	**Measured**	**ERA5**	**Measured**	**ERA5**	**Measured**	**ERA5**	**Measured**	**ERA5**
Mean	6.0	6.5	6.5	6.5	5.2	6.0	6.3	6.6
Min	3.6	4.1	4.4	4.5	2.9	3.2	4.1	4.5
Max	7.9	8.3	8.0	8.0	7.5	8.1	8.0	8.2
R	0.9910		0.9742		0.9926		0.9871	
R^2	0.9821		0.949		0.9853		0.9743	
SE	0.026		0.021		0.03		0.023	
SD	1.29		1.23		1.75		1.32	

Table 4. Monthly average of SR KWh/m^2/d for selected sites.

Months	El-Agma	El-Arish	El Golf	Habata	Sewa	Safaga	Shiab-Elbanat
JAN	**4.0**	**3.6**	**5.1**	**3.9**	**4.1**	**4.6**	**4.6**
FEB	**5.0**	**4.5**	**6.1**	**4.6**	**5.2**	**5.6**	**5.7**
MAR	**6.2**	**5.7**	**7.1**	**5.6**	**6.2**	**6.7**	**6.7**
APR	**7.2**	**6.9**	**7.7**	**6.8**	**7.1**	**7.3**	**7.5**
MAY	**7.8**	**7.7**	**7.9**	**7.8**	**8.0**	**8.0**	**8.1**
JUN	**8.0**	**7.9**	**8.0**	**8.0**	**8.1**	**8.1**	**8.2**

JUL	**7.8**	**7.6**	**8.0**	**8.1**	**8.1**	**7.8**	**7.9**
AUG	**7.3**	**7.1**	**7.6**	**7.3**	**7.5**	**7.3**	**7.5**
SEP	**6.4**	**6.2**	**7.1**	**6.4**	**6.6**	**6.7**	**6.8**
OCT	**5.4**	**5.0**	**6.2**	**4.8**	**5.4**	**5.9**	**5.9**
NOV	**4.3**	**3.7**	**5.3**	**3.9**	**4.4**	**4.8**	**4.9**
DEC	**3.6**	**3.0**	**4.7**	**3.1**	**3.6**	**4.3**	**4.3**
yearly	**6.09**	**5.74**	**6.74**	**5.84**	**6.19**	**6.42**	**6.50**
Mean	**6.10**	**5.75**	**6.75**	**5.86**	**6.20**	**6.43**	**6.51**
Standard Error	**0.08**	**0.09**	**0.06**	**0.09**	**0.09**	**0.07**	**0.07**
Standard Deviation	**1.58**	**1.76**	**1.20**	**1.80**	**1.63**	**1.35**	**1.40**
Min	**2.35**	**1.89**	**4.21**	**1.91**	**2.58**	**3.36**	**3.00**
Max	**8.25**	**8.19**	**8.44**	**8.51**	**8.42**	**8.37**	**8.45**

To see the differences in SR distribution over the Western Desert, only three stations were selected over this area, while the location and elevation of these sites are shown in Table (1). The variation of SR over these sites along the period of study is shown in figure 3. El-Golf plateau record SR is larger than the other two stations, but all three stations have the same values in the summer season. The daily values of SR over the El-Golf plateau ranged from 4.7 to 8 (KWh/m^2/d) and Habata station recorded variations of SR from 3.1 to 8.1 (KWh/m^2/d), while Sewa ranged from 3.6 to 8.1 (KWh/m^2/d). In general, the El-Golf plateau recorded more solar radiation than the others. This is due to the change in topography and its elevation from the sea level surface. Table 3 also displays the descriptive statistics for all seven stations. As can be seen from the table, El-Golf station has a higher mean for solar radiation than the other six stations, with a yearly value of 6.74 KWh/m2/d. additionally; standard error recorded a minimum value at El-Golf station and a maximum value at Sewa station, owing to the difference in topography and its lower elevation.

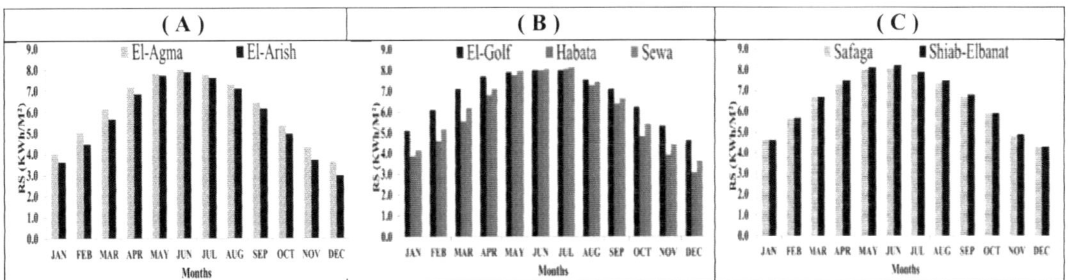

Figure 3. Monthly average Global Solar Radiation (KWh/m^2/d) for all selected sites.

The Red Sea shores of Egypt receive mild sunshine all year. The days are bright and sunny throughout the summer. However, the sea breezes should keep you cool. Warm oceans and plenty of sunshine make spring and October ideal times to visit. Winter days are still mild, but evenings can be chilly. So, two stations have been chosen in the Red Sea region to construct a PV system to support the cost of electricity and use the country's renewable energy. Figure 3 shows the result of the comparison between Safaga and Shaib-Elbanat Mountain, and there is no variation between these two stations. To see the total cost according to the PV panel in Table 2 that was used in this study over all seven sites, the formula in Eq. 1 was used to calculate the total cost. Figure 4 and Table 4 show the total cost over these seven sites. The figure illustrates that El-Arish Station has a total cost larger than El-Agma Station because its variation of solar radiation is much smaller than El-Agma Station's and finally because El-Arish is close to the Mediterranean Sea.

Table 5. Yearly average of Total cost and SR overt all stations.

Module	El-Agma	El-Arish	El-Golf	Habata	Sewa	Safaga	Shiab-Elbanat
SR (KWh/m^2/d)	6.09	5.74	6.74	5.84	6.19	6.42	6.50
M 1: η =19.1%	0.136	0.149	0.119	0.134	0.146	0.126	0.125
M 2: η=22.6%	0.165	0.180	0.143	0.162	0.176	0.152	0.151
M 3: η=22.4%	0.195	0.214	0.170	0.192	0.209	0.181	0.179

Furthermore, the station that was selected within the Western Desert is shown in figure 4. It's clear from the figure that El-Golf station has a lower cost than the other two stations because its elevation is much higher than the others. On the other hand, Sewa station has a cost larger than Habata due to its change in topography and its elevation is less than El-Golf and Habata. Figure 4 also shows the variation in the cost between Safaga and Sahib El-banat Mountain. As seen from the figure, they approximately have the same cost when compared with each other.

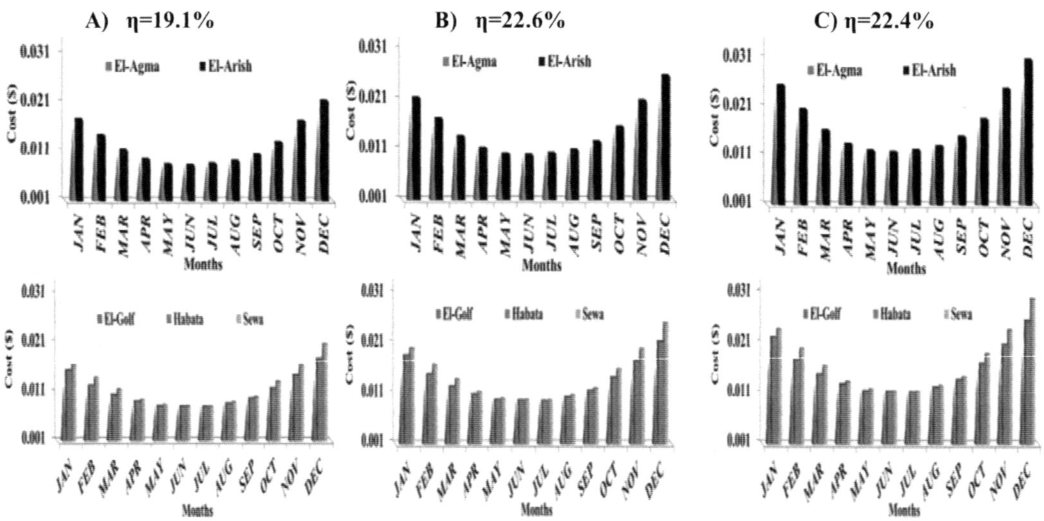

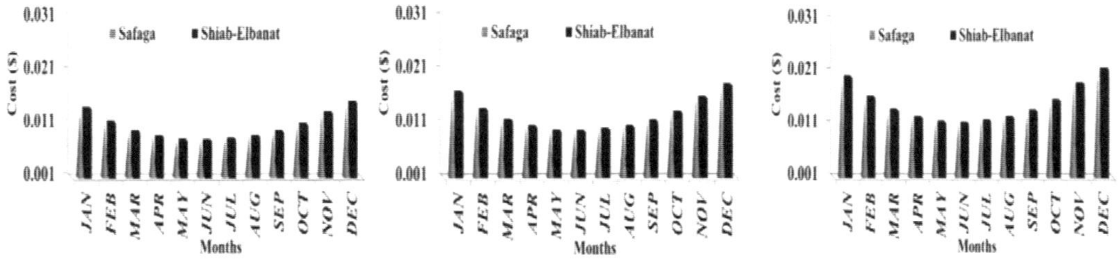

Figure 4. Total coast over Egypt with different three solar panel efficiency.

Figure 5 depicts and shows the domestic and industrial electricity consumption in Egypt; (a) In January 2022, Egypt's electricity consumption was estimated at 12688*103KWh/MN. The previous number for December 2021 was 13021*103KWh/MN. Egypt's electricity consumption data is updated monthly, with 301 observations between January 1997 and January 2022, with an average of 9578*103KWh/MN. The data peaked in Sep 2012 at 39385*103KWh/MN and hit a record low in Feb 1997 at 3737.230 KWh/MN.

(b) Egypt's electricity consumption was 7224*103KWh/MN for domestic and commercial purposes in June 2014. This represents an increase from the previous figure for May 2018 of 6239*103 KWh/MN, Egypt Electricity Consumption, Commercial and Domestic data is updated monthly, with 342 observations covering a time period from January 1990 to June 2018 with an average of 2625*103KWh/MN. The data peaked in June 2016 8420*103KWh/MN and fell to a record low in February 1990 of 954.900KWh/MN. (c) Egypt's electricity consumption for domestic and commercial purposes was 6446*103KWh/MN in June 2018. This represents an increase from the previous figure for May 2018 5609*103KWh/MN. Egypt's Domestic and Commercial Electricity Consumption Monthly updates to household data were made between July 2001 and June 2018 with 204 observations, average 3859*103 KWh/MN. The statistics peaked in June 2016 is 7644*103KWh/MN, and hit a record low in April 2002 at 2032*103KWh/MN.

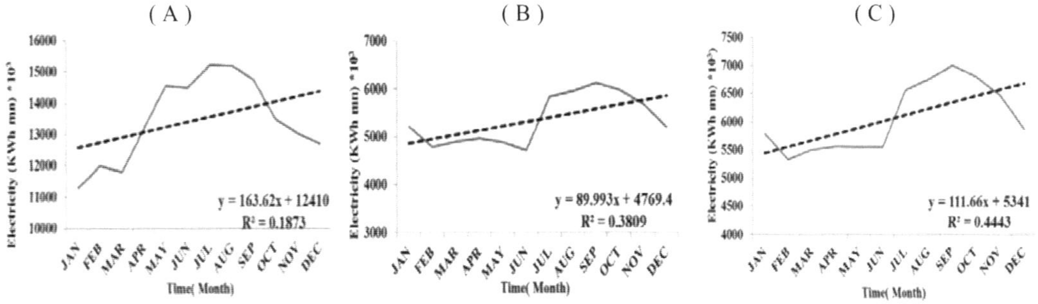

Figure 5. (a) Electricity Commercial for Egypt from Jul 1997 to jan2022, (b) Electricity household for Egypt from Jul 2001 to Dec 2017 and (c) Electricity Domestic and Commercial for Egypt from Jan 1990 to Dec 2017.

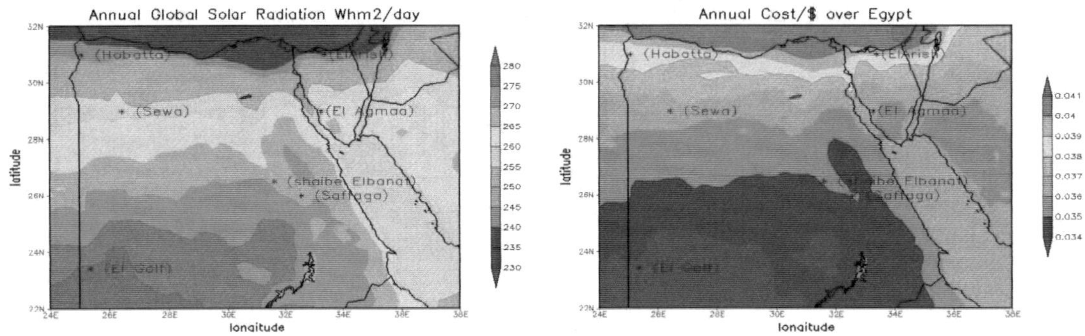

Figure 6. Average global solar radiation and the total cost over Egypt 2011-2019.

The annual (SR) and the total cost over the area of study (Egypt) is shown in Figure 6. The contour line in the figure denotes the annual total cost for the year 2017. As seen in the figure, the maximum value of (SR)is found in the south-west part of Egypt, besides while the values of the cost ranged all over Egypt from 0.03 to 0.04$.

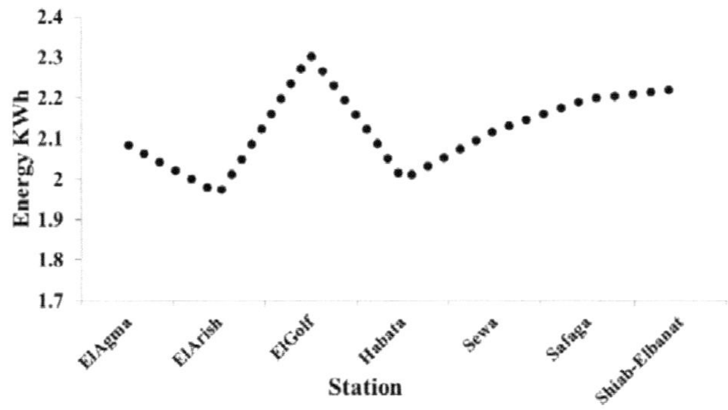

Figure 7. Annual total energy budget over the year for all seven stations (KWh/m^2/d).

6. Conclusion

This study focused on four stations and seven strategic sites distributed to cover all Egypt's territory. The observed global solar radiation (SR) data from the Egyptian Meteorological Authority (EMA) at the selected four stations (Cairo, Asyut, Kharga, and Aswan) were used to assess the accuracy of the estimated SR from the ERA5 reanalysis dataset. The other seven sites were selected carefully as strategic sites to build a PV system and produce the electricity required for national sustainability. These seven sites are characterised by their flat topography and vast areas of uninhabited land, where these entire features are great for building a PV system. Thus, the first aim of this study is to evaluate the accuracy of the ERA5 at the selected four stations against the measured SR data. Meanwhile, the second aim is to identify the best strategic sites in Egypt available to build the PV systems, in addition to calculating the total monthly and annual electricity budget and their generation costs using three PV modules (M1 (Polycrystalline), M2 (Monocrystalline), and M3 (Monocrystalline)).

It is concluded that the SR from ERA5 datasets has good agreement and reliability with the observations at the selected four stations and can be used to achieve the main objectives of this study. The performed evaluation statistics revealed that the correlation (R) is greater than 0.97 percent and the coefficient of determination (R2) exceeds 0.94 at the selected four stations. Furthermore, the slandered error (SE) and slandered deviation (SD) are small at all stations, with the SE at Asyut, Aswan, Cairo, and Kharga being 0.026, 0.021, 0.03, and 0.023, respectively, and the SD being 1.29, 1.23, 1.75, and 1.32.

Furthermore, seven strategic sites were chosen across the country, including the Western Desert (El-Golf, Habatta, and Sewa), the Eastern Desert (Saffaga, Shiab-Elbanat), and the Sinai Peninsula (El-Agma and El-Arish) to identify the most suitable station for the construction of PV systems. Three different types of PV cells with different efficiencies ($\eta 1$=19.1%, $\eta 2$=22.6% and $\eta 3$=22.4 %) were utilised to compute the energy that could be generated at the lowest cost at the seven selected sites across Egypt. The results showed that the monthly total cost from $\eta 1$ was 0.011 $/W during June and exceeded 0.015 $/W in December, while the annual total costs ranged from 0.0119 $/W in El-Golf and 0.01494 $/W in El-Arish. Also The result for cell 2 ranged from 0.01 $/W to 0.026 over El-Golf and Shiab-Elbanat, whereas cell 3 was recorded at a high cost across all sites but provides more energy. Finally, it is concluded that El-Golf, Shiab-Elbanat, and Saffaga are the best three sites with the lowest cost for building a PV system, followed by Habatta and El-Agma, while the energy generation costs at both El-Arish and Sewa were the highest.

The conclusion of this study, as shown in the previous sections, is that re-analysis data from erainteiem-ERA5 is the best in studying solar radiation, as shown in the figure above. In addition to that, the best sites for building a PV system for electricity generating are El-golf, Shiab-Elbanat, and Saffaga because they have low costs throughout the year. The annual average of (SR) for these three sites is 6.74, 6.42, and 6.5 (KWh/m^2/d) respectively, and the annual cost for M1 is 0.119, 0.126, and 0.125 $/W.

Acknowledgement

The authors gratefully acknowledge the Egyptian Meteorological Authority (EMA) for providing the observations, and the European Centre for Medium-Range Weather Forecasts (ECMWF-ERA5) for providing the estimated surface solar radiation downwards data to achieve the different purposes of our study.

Acronyms

Acronyms	Name
SR	Solar radiation
MN	month
EN	Energy
AP	Area of the panel
P	Performance ration
η	Efficiency (%)
R^2	Correlation coefficient
SE	standard error
SD	standard Deviation
CST	concentrated solar thermal
M	Module

References

[1] Behar O, Peña R, Kouro S, Kracht W, Fuentealba E, Moran L, Sbarbaro D 2021 Cleaner Engineering and Technologyt 14 100259
[2] Purohit I, Purohit P 2015 Renewable and Sustainable Energy Reviews 50 735-47
[3] Alnaser WE, Al-Kalak A, Al-Azraq MA 1995 Renewable energy 6 (5) 649-57
[4] Lorenz E, Hurka J, Heinemann D, Beyer HG 2009 IEEE Journal of selected topics in applied earth observations and remote sensing 2 (1) 2-10
[5] Lorenz E, Scheidsteger T, Hurka J. Heinemann D, Kurz C 2011 Prog.Photovoltaics Res. Appl 19, 757–771.
[6] Wang F, Mi Z, Su S, Zhao H 2012 Energies 5(5) 1355-1370
[7] Antonanzas J, Osorio N, Escobar R, Urraca R, Martinez-de-Pison FJ, Antonanzas-Torres F 2016 Solar energy 15;136:78-111
[8] Bagher AM, Vahid MM, Mohsen M 2015 American Journal of optics and Photonics 3 (5) 94-113
[9] Zempila M, Giannaros, T M, Bais A, Melas, D, & Kazantzidis A 2016 Renewable Energy 86 831-840
[10] Hoyer-Klick C, Wald L, Menard L, Blanc P, Wey E, Suri M, Cebecauer T, Huld T, Allal H, Bida A, Mahmoud M 2011 ISES Solar World Congress (pp. paper-41947)
[11] Ntwali D, Ogwang BA, Ongoma 2016 Atmospheric and Climate Sciences 6 (2) 145-57
[12] Boilley A, Wald L 2015 Renewable Energy 1 (75) 135-43
[13] Monteiro C, Santos T, Fernandez-Jimenez LA, Ramirez-Rosado IJ, Terreros-Olarte MS 2013 Energies 6 (5) 2624-43
[14] De Giorgi MG, Congedo PM, Malvoni M 2014 Measurement & Technology 8 (3):90-7
[15] Theocharides S, Makrides G, Theristis M, Kynigos M, Georghiou GE 2020 Sandia National Lab.(SNL-NM), Albuquerque, NM (United States)
[16] Ba MB, Nicholson SE, Frouin R 2001 Journal of climate 14 (1):60-76
[17] Júnior WB, Macêdo WN, Pinho JT 2014 Energy Procedia 1 (57) 4-13
[18] Del Col D, Bortolato M, Padovan A, Quaggia M 2014 Energy Procedia 1(57) 255-64
[19] IEA report 2014 input to the Clean Energy Ministerial energy technology respective 84
[20] Perez R, David M, Hoff TE, Jamaly M, Kivalov S, Kleissl J, Lauret P, Perez M 2016 Foundations and Trends® in Renewable Energy 1 (1) 1-44
[21] Čížková K, Láska K, Metelka L, Staněk M 2022 EGUsphere 7 1-32
[22] Korany M, Boraiy M, Eissa Y, Aoun Y, Abdel Wahab MM, Alfaro SC, Blanc P, El-Metwally M, Ghedira H, Hungershoefer K, Wald L 2016 Earth System Science Data 8(1) 105-13

[23] Mlawer EJ, Delamere JS, Clough SA, Miller MA, Johnson KL, Shippert TR, Long CN, Ellingson RG, Zhang MH, Ferrare RA, Cederwall RT The broadband heating rate profile (BBHRP) VAP 12th ARM Science Team Meeting Proceedings, St Petersburg, FL 2002
[24] Soulouknga MH, Falama RZ, Ajayi OO, Doka SY, Kofane TC 2017 Energy and Power Engineering 9 (12) 703
[25] Wahab MA, El-Metwally M, Hassan R, Lefèvre M, Oumbe A, Wald L 2009 International Journal of Remote Sensing (01) 261-80
[26] Denholm P, Margolis RM 2007 Energy 35 (5) 2852-61
[27] Bhandari R, Arce BE, Sessa V, Adamou R 2021 Sustainability 13 (1) 385
[28] Wolf M 1974 Photovoltaic Power Generation 49-65
[29] Maycock PD 2003 World photovoltaic markets Practical Handbook of Photovoltaics 887-912
[30] Lusardi A 2008 Financial literacy: an essential tool for informed consumer choice? National Bureau of Economic Research
[31] Aboelkhair H, Morsy M, El Afandi G 2019 Advances in Space Research 64 (1) 129-42
[32] Domroes M, El-Tantawi A 2005 A Journal of the Royal Meteorological Society 25(1) 51-63
[33] Rahoma UA 2008 American Journal of Environmental Sciences 4 (2) 121
[34] Shaltout MM, Hassan AH, Fathy AM 2001 Renewable Energy 23 (3) 605-19

IOP Conf. Series: Materials Science and Engineering 1269 (2022) 012007 doi:10.1088/1757-899X/1269/1/012007

Assessment of risk factors causing delays in road construction in Egypt

Yasser A.S Gamal[1], Elsayed Mohamed Abd Allah[2], Mina Maged[3], Mahmoud Enieb[4]

[1]Lecturer in Structural Engineering, Department of Civil Engineering, High Institute of Engineering Technology, EL-Minia, Egypt
[2]Prof. Dr. of Highways and Airports Engineering, Department of Civil Engineering, Assiut University, Assiut 71516, Egypt
[3]Master Student, Highways Engineer
[4]Assoc. Prof. of Highway Engineering, Department of Civil Engineering, Faculty of Engineering, Assiut University, Assiut 71516, Egypt

E-mail:Yasser.gamal2310@mhiet.edu.eg , emabdalla2000@gmail.com , eng.minamaged@yahoo.com , m.enieb@aun.edu.eg

Abstract Delays are a common problem in the global construction industry, affecting the development of the road construction industry, particularly in developing countries. Where, the construction industry faces a lot of risk factors that can contribute to project delays, particularly during COVID-19. The road projects risk management framework was used to identify potential risk categories that could impede the success of road construction in Egypt. On the other hand, road construction in Egypt faces numerous challenges with increasing unexpected parameters throughout the project life cycle. In other words, unanticipated risk factors have a significant impact on the success of project execution. As a result, the study's primary goals are to investigate the majority of the risk factors that cause road construction delays in Egypt. The impact of numerous risk factors has changed the delays in road projects as the COVID-19 pandemic spread across the world. As a result, the study incorporates previously unstudied factors. As well, the research has sought to investigate the new severity of risk factors during coronavirus infection. Moreover, risk breakdown structures have classified the risk factors into several categories. They include consultant risk, design risk, equipment risk, material risk, owner risk, contract risk, contractor risk, labor risk, external risk [coronavirus effects and others] and operational risk. The strategy employed in the paper is to incorporate the DEMATEL technique and risk analysis into the statistical model. Whereas the risk analysis prioritizes risk based on the evaluation of each specific project risk. Additionally, this approach has been able to create connections between risks, which may make it easier to understand the likelihood and severity of project risk than if the risks were just considered isolated events. Furthermore, the pilot survey is divided into two stages, the first of which is based on risk analysis to identify the most significant risks. As a result, the identification of the high-risk for each category has been researched. The second survey was issued to risk management by evaluating the importance of risk factors and creating causal relationships among components using the decision-making trial and evaluation laboratory (DEMATEL) approach. Finally, it is noticed that operational risk and external risk caused by the effects of the coronavirus have the greatest impact on the delay of road projects.
Keywords Road construction; Risk management; Risk breakdown structures; Qualitative risk analysis; DEMATEL.

1. Introduction

The construction industry is one of Egypt's fastest-growing sectors and plays an important role in the country's economic, societal, and political development [1]. Delays are a common problem in the global construction industry, affecting the development of the road construction industry, particularly in developing countries. Furthermore, the construction industry faces a lot of risk factors that can contribute to project delays, particularly during COVID-19. On the other hand, the project risks take into account a measurable aspect by estimating both probability and impact. It can also be a threat or an opportunity. Furthermore, the risk measurements show the deviation from the project's planned trend. As a result, risk analysis is an important process from project selection to construction work execution. In other words, risk management has the potential to influence project success, which can be critical in the construction industry. Furthermore, an effective risk management process has been a key pillar in meeting time, cost, quality, scope, and other objectives. In addition, construction delays are a common major impediment to the completion of road projects. The Egyptian road sector has focused on project delays and cost overruns. So, the primary goal of a construction project is to meet the functional requirements of the owner while adhering to the project's schedule, budget, and quality objectives [2]. Also, Contractors' failure to deal with unanticipated risks frequently leads to failure to achieve such goals. The PMBOK presented a definition of risk as an unanticipated occurrence that has a negative or positive impact on the project's objectives. Risk management, as a subset of project management, encompasses the processes of risk management planning, identification, analysis, response planning, and risk control on a project [3]. Effective risk management necessitates creative risk identification, evaluation, and control to convert as many unknown risks as possible into known risks. In general, the goal of risk management is to increase opportunities [positive events] while decreasing threats [negative events][4]. Furthermore, inaccuracies in risk management result in time and cost overruns [5]. Although Egypt's construction industry has grown economically, it has been subjected to numerous risks and uncertainties, resulting in an inability to meet project objectives. As well, it can lead to inaccurate time and cost estimates. And, it can also be caused by the lack of a comprehensive risk management approach that can anticipate, analyze, create a response plan, and manage risks. Therefore, the study's primary goals are to investigate the majority of the risk factors those causing road construction delays in Egypt. In addition, the study included a plethora of risk factors to demonstrate the severity of its impact, particularly during COVID-19. Moreover, risk breakdown structures have classified the risk factors into several categories. They include consultant risk, design risk, equipment risk, material risk, owner risk, contract risk, contractor risk, labor risk, external risk [coronavirus effects and others] and operational risk. Furthermore, the pilot survey is divided into two stages, the first of which is based on risk analysis to identify the most significant risks. As a result, the identification of the high-risk for each category has been researched. The second survey was issued to risk management by evaluating the importance of risk factors and creating causal relationships among components using the decision-making trial and evaluation laboratory (DEMATEL) approach.

2. Literature Review

Construction projects all over the world involve numerous risks. As a result, things can go as planned; risks in construction projects are practically expected. In addition, cost and time overruns are common in construction projects. As well, the delay is viewed as an occurrence or disruption that causes the project to be completed later than the contract stipulates. In other words, a delay is defined as a date that is later than the planned start or end date. Furthermore, the time delay could be described as an event or disorder that results in the project being completed after the contract's stipulated duration. A delay could also be caused by starting or finishing an activity later than planned. In other words, a delay is an action or activity that increases the time required to carry out contract activities [6]. According to the project management institute's PMBOK® Guide, the definition of risk in any project is an uncertain event or condition that, if it occurs, has a positive or negative impact on one or more project objectives such as scope, schedule, cost, and quality. A risk may have one or more causes and, if it occurs, one or more consequences. A cause can be an assumption, a constraint, circumstances, a given requirement, or a potential requirement that results in the possibility of a negative or positive effect [7]. In other words, it is the possibility of losing or gaining money. Furthermore, the risk is anything that could jeopardize the project's schedule, cost, quality, or scope. Moreover, most organizations' established project management procedures do not accommodate the growing demands for risk management, and as a result, many projects are not set up to manage risk [8]. Furthermore, many organizations believe that project management focuses solely on time and cost as the primary objectives for project success. Either risk may have an impact on at least one of the project constraints, such as cost, time, or quality. However, time risks are both threats and opportunities. And, threats tend to reduce the effectiveness of achieving the company's objectives and goals. Furthermore, performance is measured against objectives that are set far in advance of the project execution stage but are frequently based on incomplete information [8]. As mentioned above, project risk management refers to the processes involved in risk management planning, identification, analysis, responses, and project monitoring and control. Many researchers investigated the causes of risks in the construction industry in order to provide stakeholders with tools for identifying risks. Ehsan and Mirza demonstrated construction industry risk factors such as time crunch, resource availability, history, design complexity, experience, management steadiness, and team size. They also classified construction risks as technical, logistical, management, environmental, financial, and sociopolitical [9]. Odabaş investigated the risk factors that influence construction period length and estimated the construction work period. The researcher selected the most important publications and classified them into eleven categories [10]. Enshassi and Mayer identified a list of risk factors for construction projects and classified them into seven categories. And, the categories of risk include physical, political, financial, environmental, design, and site construction risks [11]. Hosny, Ibrahim compiled a comprehensive list of risks associated with the CFA [continued flight of Auger piles] in Egypt [12]. As a result, the qualitative analysis assigned to the main CFA risk factors [design, equipment, economics, construction, management, owner, political, and subcontractor risks] is 16.5, 15.1, 14.3, 11.6, 11.3, 7.4, 4.5, and 4.1 percent, respectively [12]. Mubarak classified the causes of construction delays into six categories, which are as follows: manager changes, design flaws or omissions, inclement weather, other conditions, and force majeure requirements. [13]. On the other hand, Tang and Qiang determined 32 risk factors for construction projects in China and sorted them into five categories: design, quality, safety, facility, and financial [14]. In addition, Kuo and Lu presented twenty-five risk factors for construction projects to improve project efficiency [15]. The main cause of construction project delays was identified using fault tree analysis by Yasser and Mostafa [16]. Also, they created a conceptual framework of delay factors influencing Egyptian construction project completion dates [17].

On the other hand, individuals and organizations can define decision-making as selecting the best option for achieving their goals within the constraints of the current situation. Furthermore, decision-making is an interdisciplinary topic of study that attracts researchers and academics from a variety of fields. Also, experience and scientific approaches are important tools for decision-making. Previous studies proposed various methodologies for risk assessment in construction projects. In recent years, new approaches based on various methodologies have been described. MCDM approaches (Multi-criteria decision-making methods) have been developed for the proper evaluation of multiple different criteria in solving complicated situations[18]. Furthermore, the increasing interest in MCDM methods has resulted in a steady increase in the publication of the DEMATEL method. Many studies have been conducted to examine risks in the construction industry using MCDM methods[19],[20]. Boateng and Chen suggested a model for risk ranking in transportation megaprojects by combining Analytic Network Process (ANP) and risk priority index methodologies [21]. The Battelle Memorial Institute's Geneva Research Centre developed the DEMATEL (Decision Making Trial and Evaluation Laboratory) technique to depict the structure of complex causal interactions using matrices or digraphs [22]. It's particularly beneficial in assessing the cause-and-effect relationships among risk factors. Therefore, it is considered a type of structural modeling approach. The DEMATEL can be used to investigate and solve complicated and intertwined problems by confirming interdependence among factors and assisting in the development of a map to reflect relative relationships within them. In the other words, the fundamental aim was to create a structural model utilizing a scale to score the elements through pair-wise comparisons to depict the cause-and-effect relationships between risk factors and their influence on the system. DEMATEL's technique has received a lot of attention in the last decade due to its benefits and capabilities, and many academics have used it to solve complex system issues in a variety of fields [23],[24],[25],[26]. Furthermore, although global systems contain imprecise and uncertain information, the DEMATEL has been developed to aid in decision-making in a variety of situations.

The goal of the research is to develop a framework for managing the risk factors that cause delays in Egyptian road projects, allowing organizations to follow the steps outlined below:

- Identification of the risk factors,
- Qualitative risk analysis,
- Evaluating the importance of risk factors and creating causal relationships among components using the decision-making trial and evaluation laboratory (DEMATEL) approach.

The proposed framework assists in precisely anticipating risk events and assessing their characteristics, allowing for the selection of appropriate proactive actions with accurate time contingency reserves. And, it allows for a smooth project schedule.

3. Research Methodology

The proposed study attempts to elicit the specific risk with as much expert judgment as possible through interviews and pre-structured questionnaires, analyzing each identified risk. It is possible to accomplish this by doing the following:

- Creating a quick list of the most common risks facing road projects in Egypt using a brainstorming-directed questionnaire based on specialist inputs demonstrating the causes, risk events, and impacts for each identified risk and its probability.
- Classification of the risk factors into several types based on their type and nature. Hence, it creates the risk of breakdown structures [RBS], which is part of the risk management plan.
- Prioritizing identified risks based on risk score estimates produced by evaluating and multiplying the likelihood of occurrence and time impact.
- Creating causal relationships among components using the decision-making trial and evaluation laboratory (DEMATEL) approach.

Additionally, Figure 1 illustrates the integration of the qualitative risk analysis and the DEMATEL approach. The qualitative risk analysis was utilized to determine the high-risk factors for each category and all risk factors, as indicated in the process figure. Sequence, the DEMATEL approach was employed to establish the causal links between the risk factors as shown in Figure 1.

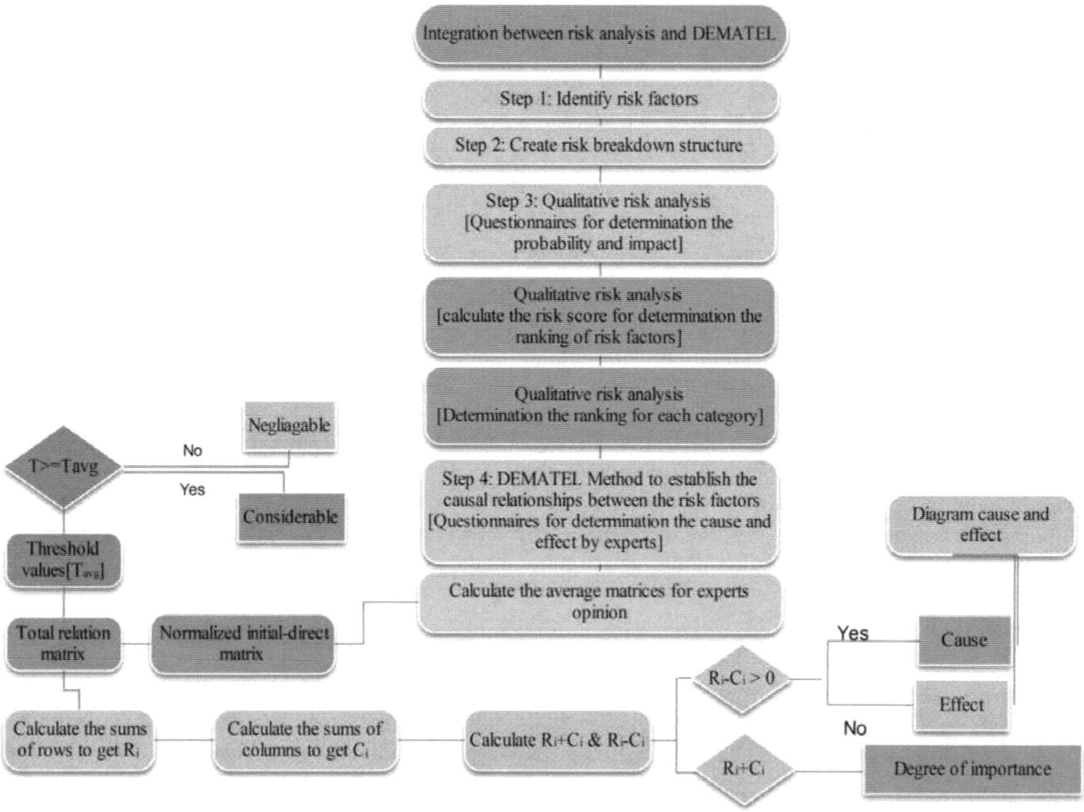

Figure 1. Flow chart for the process of the integration between qualitative risk analysis and the DEMATEL method

4. Risk Management Plan

It includes deciding how to prepare for dealing with the risk and treatment entrances, as well as identifying actions, coordination, and planning processes. This is very significant because it demonstrates the next steps to be taken for the identification of the severity, type, and extent of risk in the project, as well as the appropriate treatment [7]. The first step of the plan risk management is the meeting for diagnosis risk, determining the methods for qualitative and quantitative risk analysis, as well as procedures for response and controlling the risk. And, the risk category is a part of the risk management plan. Where risk breakdown structure [RBS] can classify the risk according to its nature and type. Also, it plots a hierarchical chart that decomposes the project risks, starting with higher-level categories and continuing down into sub-levels of risk. Therefore, the study classified the risks associated with road projects in Egypt into eleven categories. It includes consultant risk, design risk, equipment risk, material risk, owner risk, contract risk, contractor risk, labor risk, external risk [coronavirus effects and others] and operational risk. This is a step before the identification of risk, which can help determine the comprehensive risks of the construction project. Figure 2 shows the classification of the risk type by the risk of breakdown structure.

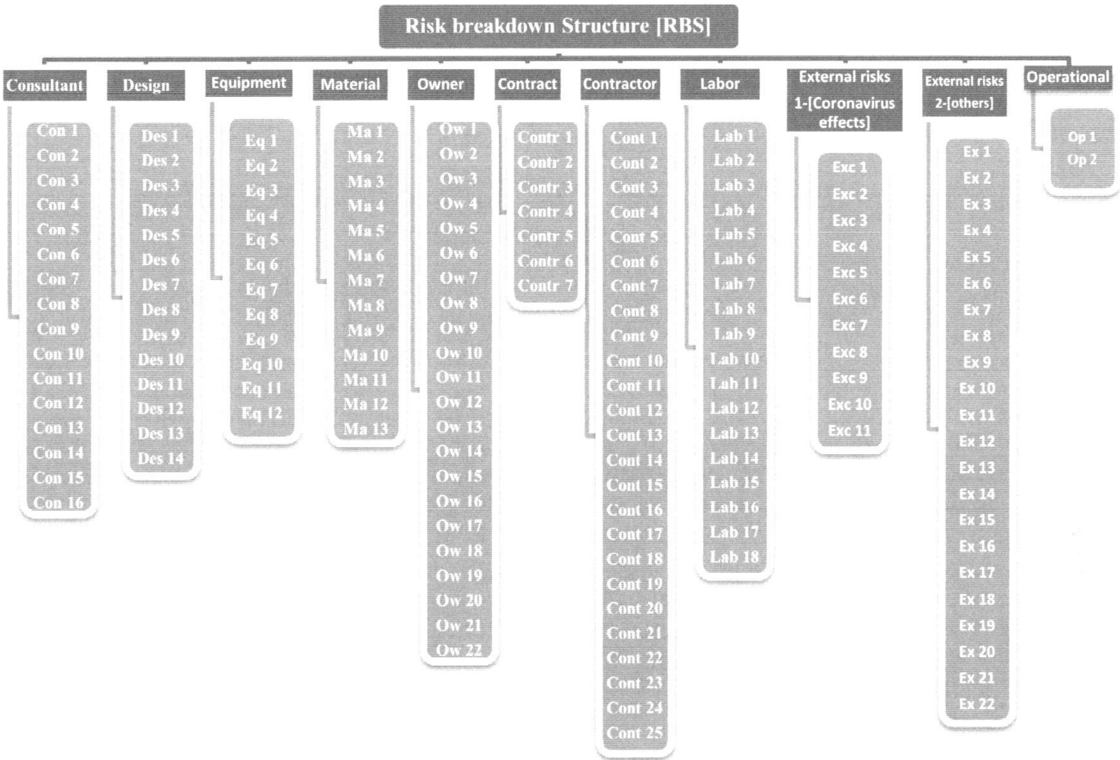

Figure 2. Risk Breakdown Structure

5. Risk Identification

The goal of risk identification is to create a comprehensive list of potential risks associated with the construction of roads in Egypt. Moreover, there are many tools and techniques to determine the risk, such as literature review, brainstorming, interviewing, meeting with experts, and creating a checklist that can be created based on historical data and knowledge from previous projects. So, the research depends on different techniques to get the list of risks with different categories. And, the comprehensive list of risks has been created based on previous experience, knowledge, and lessons learned from similar projects that have been completed. In addition, the study included a plethora of risk factors to demonstrate the severity of its impact, particularly during COVID-19. As a result of coronavirus effects, new factors have been involved that have not previously been studied.

6. Data Collection

The data collection and questionnaire design processes begin with the creation of a sample questionnaire. A pilot study was conducted on firms using an interview and a questionnaire to identify the risks causing road project delays in Egypt. The data was gathered from 60 knowledgeable participants (see Table 1).

IOP Conf. Series: Materials Science and Engineering 1269 (2022) 012007 doi:10.1088/1757-899X/1269/1/012007

Table 1. Expert's Profile, the data was gathered from 60 knowledgeable participants

No / Percentage	Experience [years]			
	Under 5	Between 5 to 10	Between 11 to 15	Above 15
No	7	35	9	9
Percentage	11.67 %	58.33 %	15.00 %	15.00 %
No / Percentage	Type of work			
	Consultant	Contractor	Owner	
No	17	35	8	
Percentage	28.33 %	58.33 %	13.33 %	
No / Percentage	Type of sector			
	Government	Private		
No	30	30		
Percentage	50.00 %	50.00 %		

7. Qualitative Risk Analysis

The Perform qualitative risk analysis process analyzes and evaluates the features of individually identifiable project risks, then prioritizes them based on their attributes. In addition, individual risks are assessed using qualitative risk analysis, which determines the likelihood of each risk occurring and the impact of each risk on the project's objectives [7]. Its goal is to compute the probability of occurrence (P) and impact of the identified risks (I) to rank and categorize them for further investigation or action. Furthermore, this might assist the project manager in focusing on high-priority risks that do require action. In other words, high-priority risks in terms of threats are considered a major focus of the plan risk response process. Therefore, the following steps should be included in the analytical process:

- Defining a subjective risk probability (P) value on a scale ranging from 0 to 1 (e.g. 0.1, 0.3, 0.5, 0.7, 0.9).
- Establishing a subjective risk impact (I) on a scale using the standard impact evaluation method for schedule impacts is shown in Table.2.
- The value of probability can be determined by getting the total probabilities for each risk from the questionnaires and hence finding the mean. And also, the impact can be identified by the same method.
- Multiplying the probability (P) by the impact (I) to get the risk score (RS), where RS can be calculated by Equation (1)

$$RS = P \times I \qquad (1)$$

- Risks are prioritized in descending order depending on their risk score values, and risk ratings are recognized (RR).

Sorting the risk categories (RC) into high (H), moderate (M), or low (L) risks using the probability and impact matrix as shown in Table 3. Then, high and moderate risks have been chosen (risks that merit a response).

Table 2. Standard of risk impact [7]

Impact Scales

Project Objectives	Relative or numerical scales are shown				
	Very low /0.05	Low /0.10	Moderate /0.20	High /0.40	Very high /0.80
	Probability Scales				
	Very low /0.1	Low /0.30	Moderate /0.50	High /0.70	Very high /0.90
Time	Insignificant time	<5% time increase	5-10% time increase	10-20% time increase	> 20% time increase

Table 3. Probability and impact matrix[7]

Probability and Impact Matrix										
P	**Threats**					**Opportunities**				
0.90	**0.05**	**0.09**	**0.18**	**0.36**	**0.72**	**0.72**	**0.36**	**0.18**	**0.09**	**0.05**
0.70	**0.04**	**0.07**	**0.14**	**0.28**	**0.56**	**0.56**	**0.28**	**0.14**	**0.07**	**0.04**
0.50	**0.03**	**0.05**	**0.10**	**0.20**	**0.40**	**0.40**	**0.20**	**0.10**	**0.05**	**0.03**
0.30	**0.02**	**0.03**	**0.06**	**0.12**	**0.24**	**0.24**	**0.12**	**0.06**	**0.03**	**0.02**
0.1	**0.01**	**0.01**	**0.02**	**0.04**	**0.08**	**0.08**	**0.04**	**0.02**	**0.01**	**0.01**
I	**Very low /0.05**	**Low /0.10**	**Moderate /0.20**	**High /0.40**	**Very high /0.80**	**Very high /0.80**	**High /0.40**	**Moderate /0.20**	**Low /0.10**	**Very low /0.05**

As a result, the questionnaire's clarity, completeness, and applicability are confirmed. And, eleven main categories are divided into the risk factors. Moreover, each category is subdivided into several risk factors as follows in Table 4.

The risk score (RS) has been used to determine the risk ranking (RR). However, each main risk category's total risk score (TRS) has been calculated as the sum of all risk scores within that category divided by number (n) of risk factors in each category, as indicated in Equation (2).

$$TRS = \sum RS / n \qquad (2)$$

The following risk categories were prioritized in descending order as follows: operational risk, external risk [coronavirus effects], equipment risk, external risk [others], contract risk, contractor risk, design risk, material risk, owner risks, labor risk and consultant risk, with TRS of 0.187, 0.161, 0.155, 0.145, 0.124,0.120, 0.119, 0.113, 0.110, 0.094 and 0.091, respectively, as shown in Figure 3. As a result, the operational risk and the external risk caused by the effects of the coronavirus have the greatest impact on the delay of road projects.

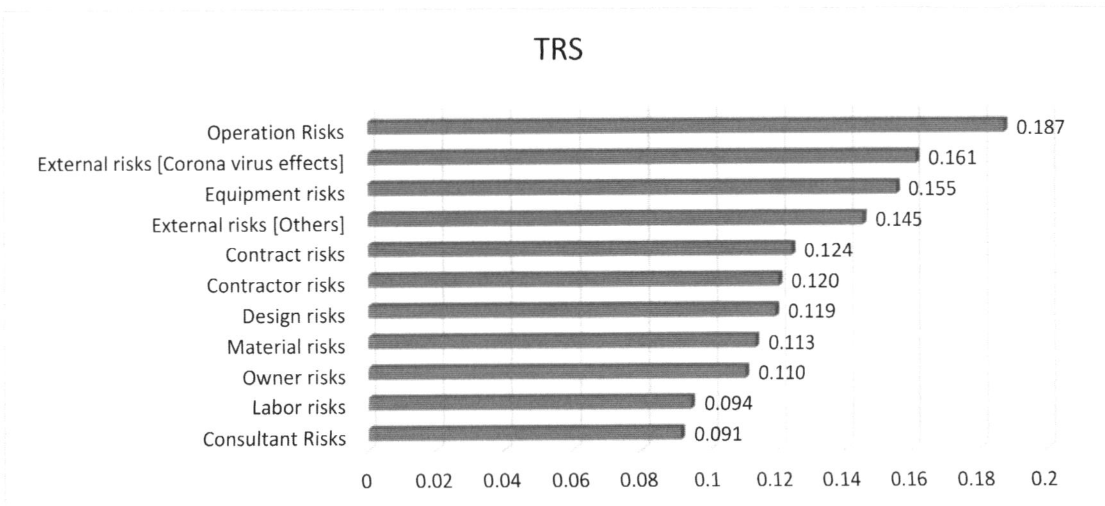

Figure 3. TRS

Table 4. Analysis of risk factors

Risk Code	Risk Factor	P	I	RS	Risk Case	(RRC) Risk Ranking with respect to category	(RRT) Risk Ranking for total risk factors
	Consultant Risk				TRS = 0.091		
Con1	The interference of the owner in the consultant's decisions	0.45	0.25	0.113	Moderate	3	89
Con 2	Conflicts between consultant and design	0.38	0.28	0.105	Moderate	4	93
Con 3	The slow reply to questionnaires by the contractor	0.32	0.28	0.090	Moderate	8	125
Con 4	Insufficient experience in the supervision crew	0.43	0.33	0.141	Moderate	1	50
Con 5	Inflexibility of supervision	0.44	0.31	0.136	Moderate	2	58
Con 6	Weakness of documented systems in the consultant office	0.36	0.24	0.087	Moderate	9	129
Con 7	Absence of Quality Assurance / Control	0.28	0.22	0.062	Moderate	14	158
Con 8	Temporary existence of the supervision crew in the project	0.21	0.22	0.046	Low	16	160
Con 9	Failure to adopt and adhere to the project schedule	0.31	0.28	0.084	Moderate	10	132
Con 10	Delay in approval of changes during the execution of work	0.32	0.28	0.091	Moderate	7	123
Con 11	Delay in reviewing and approving the design documents	0.32	0.32	0.102	Moderate	5	103
Con 12	Delay in approving the finished work	0.29	0.34	0.099	Moderate	6	107
Con 13	Delay in approving the material samples	0.27	0.28	0.078	Moderate	12	145
Con 14	Soil investigation is inaccurate	0.26	0.30	0.077	Moderate	13	146
Con 15	The lack of continuously updating the schedule	0.32	0.25	0.081	Moderate	11	140
Con 16	Absence of the notice date to proceed	0.28	0.21	0.060	Low	15	159
	Design risk				TRS = 0.119		
Des 1	Misunderstanding of the owner's requirements	0.29	0.30	0.088	Moderate	14	127
Des 2	The owner's continuous intervention and his modifications of the designer's decisions	0.48	0.36	0.174	Moderate	2	18
Des 3	Lack of procedures for information collection and topographic survey	0.35	0.34	0.118	Moderate	5	82
Des 4	The delay in approving the design of hydrological studies (studies to know the discharge of rainwater and torrential rain)	0.32	0.28	0.090	Moderate	13	126
Des 5	Delay in receiving the design documents	0.33	0.31	0.103	Moderate	9	99

Risk Code	Risk Factor	P	I	RS	Risk Case	(RRC)	(RRT)
Des 6	Delay in reply to the questions submitted by the contractor	0.32	0.29	0.091	Moderate	12	124
Des 7	The details of the drawings are unclear	0.36	0.25	0.091	Moderate	11	122
Des 8	Absence of constructability analysis	0.36	0.27	0.098	Moderate	10	109
Des 9	The schedule doesn't involve the start and end dates of the design.	0.37	0.28	0.104	Moderate	8	96
Des 10	Designs without considering the effects of environmental factors such as floods and rain	0.37	0.31	0.116	Moderate	6	85
Des 11	The design doesn't finish at the beginning of the project	0.52	0.39	0.201	High	1	7
Des 12	Conflicts between drawings and specifications	0.34	0.32	0.109	Moderate	7	91
Des 13	Inappropriate design (such as not considering the nature of the quarries in the area, and consequently the contractor's inability to supply the required materials - etc.)	0.43	0.36	0.152	Moderate	3	37
Des 14	Failure to design alternative paths to the road being implemented to transport traffic on these paths during the implementation process leads to obstruction of road implementation and many accidents on the road being implemented	0.44	0.31	0.137	Moderate	4	56
	Equipment risk			TRS = 0.155			
Eq 1	Frequent equipment failures and lack of spare parts	0.44	0.41	0.180	High	4	13
Eq 2	Low both efficiency and productivity of equipment	0.43	0.36	0.157	Moderate	7	32
Eq 3	Delays in fixing the equipment	0.48	0.40	0.192	High	2	9
Eq 4	The company doesn't study the technical status of equipment and the proper distribution of the equipment according to the work requirements.	0.45	0.35	0.158	Moderate	6	31
Eq 5	Equipment shortage	0.47	0.41	0.192	High	3	10
Eq 6	Difficulty moving equipment, whether for periodic maintenance or working on another part of the project.	0.42	0.30	0.126	Moderate	9	71
Eq 7	the careless of getting the license or renewal of the license of the equipment	0.40	0.26	0.104	Moderate	12	97
Eq 8	Lack of high-technology mechanical equipment	0.52	0.32	0.164	Moderate	5	25

Risk Code	Risk Factor	P	I	RS	Risk Case	(RRC)	(RRT)
Eq 9	The distance between the asphalt mixer location and the crushers from the project is considerable.	0.37	0.33	0.124	Moderate	10	74
Eq 10	Lack of skill of technicians and engineers in asphalt mixers and crushers (lack of interest in pre-work tests; lack of interest in training technicians; insufficient experience)	0.40	0.31	0.123	Moderate	11	75
Eq 11	Equipment allocation problem	0.44	0.30	0.133	Moderate	8	61
Eq 12	Using old equipment with low rates, with no attention to periodic maintenance	0.51	0.41	0.211	High	1	6
Material risk					**TRS= 0.113**		
Ma 1	Vendor materials do not meet specifications	0.42	0.31	0.130	Moderate	4	63
Ma 2	Unreliable suppliers	0.35	0.30	0.105	Moderate	8	94
Ma 3	Materials damage	0.28	0.25	0.072	Moderate	13	149
Ma 4	Delay in preparing the material request	0.34	0.33	0.114	Moderate	6	87
Ma 5	Changes in material types and specifications during construction	0.38	0.26	0.101	Moderate	9	106
Ma 6	Escalation of prices in material	0.53	0.33	0.176	Moderate	1	15
Ma 7	Weak strategies for procurement of construction material	0.40	0.29	0.119	Moderate	5	81
Ma 8	Use of low-quality materials	0.37	0.25	0.093	Moderate	10	120
Ma 9	Lack of production of the base layer or materials used in asphalt	0.41	0.35	0.143	Moderate	2	47
Ma 10	The delivery of the materials to the working area is difficult because the location of the project is remote from the fabrication and supply area	0.41	0.35	0.142	Moderate	3	49
Ma 11	The presence of mud in the quarries and the failure to develop successful solutions to mud treatment during implementation	0.36	0.32	0.113	Moderate	7	88
Ma 12	The weakness in the quality control	0.35	0.24	0.084	Moderate	12	136
Ma 13	The RC3000 or MCO adhesive is law quality impregnation or the percentage of bitumen in the mixture is high or the temperature of the air can lead to asphalt leaching	0.36	0.23	0.084	Moderate	11	134
Owner risk					**TRS = 0.110**		
Ow 1	The financial situation of the owner is weak	0.26	0.36	0.093	Moderate	13	121
Ow 2	The delay in the progress payment	0.41	0.35	0.144	Moderate	4	46
Ow 3	Lack of incentives for the contractor to finish early	0.35	0.23	0.080	Moderate	18	141

Risk Code	Risk Factor	P	I	RS	Risk Case	(RRC)	(RRT)
Ow 4	Tender award for the lowest price regardless of the contractor's competence to complete the project on time and with the required quality	0.51	0.36	0.184	High	1	12
Ow 5	The schedule according to the contract conditions is not realistic	0.42	0.29	0.122	Moderate	9	76
Ow 6	Ineffective delay penalties	0.28	0.26	0.071	Moderate	20	150
Ow 7	Weakness of the owner's strictness regarding the contract conditions	0.29	0.28	0.082	Moderate	17	139
Ow 8	Increasing the quantities of execution works due to a shortage of studies before starting work	0.34	0.26	0.087	Moderate	14	130
Ow 9	Increasing the modifications of the project plans during implementation due to shortcomings in the study	0.45	0.36	0.160	Moderate	2	29
Ow 10	Frequent changes in project managers	0.33	0.21	0.069	Moderate	21	152
Ow 11	Frequent change orders by the owner during execution	0.42	0.33	0.141	Moderate	5	51
Ow 12	Adding items or work not stipulated in the contract leads to scope creep.	0.43	0.35	0.152	Moderate	3	38
Ow 13	Poor communication between the owner and other partners	0.26	0.25	0.066	Moderate	22	155
Ow 14	The slow resolution of the conflict between the execution companies	0.28	0.30	0.084	Moderate	16	135
Ow 15	The slowness of making a decision	0.30	0.29	0.086	Moderate	15	131
Ow 16	Lack of leadership skills of the owner	0.34	0.29	0.097	Moderate	12	111
Ow 17	Excessive bureaucracy in managing the project and imposing the opinion of the owner	0.45	0.28	0.127	Moderate	8	69
Ow 18	Bias or unfair treatment policies between the executing companies of the owner	0.39	0.27	0.102	Moderate	11	102
Ow 19	Weakness in solving the obstacles of the executing company, such as the difficulty of providing bitumen when the financial dues are delayed	0.43	0.32	0.137	Moderate	7	55
Ow 20	Weakness of the supervisory side by the owner	0.35	0.23	0.079	Moderate	19	143
Ow 21	Postponing the delivery of documents by the owner (design-correspondence, etc.)	0.42	0.33	0.140	Moderate	6	52
Ow 22	Delay in handing over the site without obstacles from the owner to the contractor	0.33	0.35	0.115	Moderate	10	86

Risk Code	Risk Factor	P	I	RS	Risk Case	(RRC)	(RRT)
Contract risk		TRS = 0.124					
Contr 1	Lowering the value of penalties on the contractor in the event of a breach of the contract.	0.29	0.26	0.075	Moderate	6	148
Contr 2	Weakness of the equation for price differences (unfairness of price differences)	0.34	0.21	0.071	Moderate	7	151
Contr 3	Request for additional work out of the project scope	0.46	0.36	0.164	Moderate	2	26
Contr 4	Changes in contract quantities	0.48	0.36	0.173	Moderate	1	20
Contr 5	Starting the implementation of the work by assigning the contractor without knowing the prices and quantities of the items	0.47	0.27	0.125	Moderate	4	73
Contr 6	Quantities above 125% of the contract	0.42	0.28	0.118	Moderate	5	83
Contr 7	Delayed disbursement of price differences to subcontractors	0.51	0.28	0.145	Moderate	3	45
Contractor Risk		TRS= 0.120					
Cont1	Difficulties in the financial funding of the contractor	0.51	0.41	0.212	High	1	5
Cont 2	Poor Managerial Skills	0.41	0.31	0.129	Moderate	10	66
Cont 3	Inadequate contractor experience	0.40	0.33	0.129	Moderate	9	65
Cont 4	Improper pricing by contractors to win the bid	0.36	0.29	0.104	Moderate	17	98
Cont 5	Breach of contract by the contractor	0.36	0.27	0.096	Moderate	21	115
Cont 6	Shortage of technical personnel	0.39	0.30	0.120	Moderate	13	80
Cont 7	Lack of project management office (PMO)	0.52	0.29	0.148	Moderate	3	41
Cont 8	The weakness of training and development of engineers and engineering departments	0.52	0.26	0.133	Moderate	7	60
Cont 9	Inadequate contractor review and poor pre-handover inspection	0.41	0.24	0.099	Moderate	19	108
Cont 10	Rework due to the improper execution of work	0.40	0.26	0.106	Moderate	15	92
Cont 11	Mismanagement of the project schedule	0.43	0.30	0.129	Moderate	8	64
Cont 12	Failure to develop a plan for priorities according to the schedule	0.43	0.30	0.128	Moderate	12	68
Cont 13	Not estimating productivity, not following up on the daily rates of implementation, and not comparing them with what is required according to the plan	0.33	0.29	0.094	Moderate	23	118
Cont 14	The delay in preparing shop drawings and material samples by a contractor	0.34	0.25	0.087	Moderate	24	128
Cont 15	Inappropriate construction methods	0.43	0.32	0.138	Moderate	6	54
Cont 16	Lack of High-Technology	0.48	0.29	0.142	Moderate	5	48

IOP Conf. Series: Materials Science and Engineering 1269 (2022) 012007 doi:10.1088/1757-899X/1269/1/012007

Risk Code	Risk Factor	P	I	RS	Risk Case	(RRC)	(RRT)
Cont 17	Delay in resource mobilization	0.39	0.32	0.128	Moderate	11	67
Cont 18	The absence of following the rules and instructions on safety inside the project	0.50	0.16	0.079	Moderate	25	142
Cont 19	Disadvantages of night work and overtime work (poor quality - work fatigue, lack of workers incentives)	0.45	0.22	0.101	Moderate	18	105
Cont 20	Conflicts between sub-contractors during the execution of the schedule plan and also the absence of coordination between them to commit the sequence of the planned schedule	0.38	0.28	0.104	Moderate	16	95
Cont 21	Lack of coordination between subcontractors on site	0.35	0.27	0.097	Moderate	20	113
Cont 22	Delay in works of subcontractor	0.40	0.39	0.154	Moderate	2	35
Cont 23	Unreliable subcontractor	0.34	0.28	0.095	Moderate	22	116
Cont 24	Frequent change of sub-contractors because of their inefficient work	0.40	0.37	0.148	Moderate	4	42
Cont 25	The change frequently sub-contractors because of their inefficient work.	0.38	0.29	0.112	Moderate	14	90
	Labor risk				**TRS= 0.094**		
Lab 1	Shortage of labor for work, especially skilled labor for the contractor (engineers, technicians, equipment operators, labor trained on surveying equipment, etc.).	0.42	0.34	0.146	Moderate	2	44
Lab 2	Labor, particularly highly qualified workers, technicians, or engineers, is lost as a result of the resignation.	0.46	0.33	0.153	Moderate	1	36
Lab 3	Labor Dispute, strikes and personal conflicts among labor	0.31	0.22	0.068	Moderate	13	153
Lab 4	The conflict between the workers and the staff team	0.31	0.21	0.065	Moderate	15	156
Lab 5	Lack of affiliation and sincerity at work	0.32	0.23	0.076	Moderate	12	147
Lab 6	Behavioral risks that are incompatible with work ethics	0.25	0.16	0.041	Low	18	162
Lab 7	Changes in workers' wages and delays in workers' salaries	0.38	0.27	0.102	Moderate	8	101
Lab 8	Lack of incentive for workers to finish work according to plan	0.47	0.29	0.137	Moderate	3	57
Lab 9	Low morale of workers	0.42	0.29	0.122	Moderate	4	77
Lab 10	Low productivity of labor	0.40	0.30	0.120	Moderate	5	79
Lab 11	The lack of tests to select the appropriate workforce to work on the project	0.42	0.28	0.117	Moderate	6	84
Lab 12	Discriminatory practices among project workers	0.44	0.23	0.103	Moderate	7	100
Lab 13	The conflict between contrary decisions in the work	0.40	0.24	0.095	Moderate	9	117

Risk Code	Risk Factor	P	I	RS	Risk Case	(RRC)	(RRT)
Lab 14	The delay in the arrival of laborers to the work region	0.32	0.27	0.084	Moderate	10	133
Lab 15	Labor accidents and injuries	0.25	0.17	0.043	Low	17	161
Lab 16	The medical issues for labor	0.33	0.20	0.067	Moderate	14	154
Lab 17	The increasing number of working hours can lead to work fatigue.	0.42	0.20	0.083	Moderate	11	138
Lab 18	Frequent absenteeism	0.30	0.20	0.062	Moderate	16	157
External risk [coronavirus effects]		**TRS=0.161**					
Extc 1	The effects of coronavirus on the cessation and lowering of productivity	0.40	0.37	0.149	Moderate	8	40
Extc 2	Many projects have been suspended due to the coronavirus pandemic.	0.44	0.39	0.173	Moderate	4	19
Extc 3	Reducing employment to 50%	0.46	0.38	0.175	Moderate	3	17
Extc 4	Continuous changes in legislation by the government	0.45	0.36	0.161	Moderate	6	28
Extc 5	Delaying and stopping off the supplies inside and outside Egypt due to the closure of the countries	0.52	0.44	0.228	High	1	2
Extc 6	Difficulty in carrying out the tasks of work in light of the coronavirus pandemic	0.46	0.33	0.149	Moderate	7	39
Extc 7	The frequent absence of labor, whether due to health conditions, the difficulty of transportation, or fear of a fine for breaking the curfew, etc. (during the coronavirus pandemic)	0.47	0.34	0.161	Moderate	5	27
Extc 8	Low morale and poor psychological state among workers due to the isolation of the work site due to the coronavirus pandemic	0.43	0.30	0.126	Moderate	11	72
Extc 9	Weakness and lack of interest of workers in preventive measures against the Coronavirus and the spread of infection	0.57	0.24	0.136	Moderate	9	59
Extc 10	Slow decision-making by the owner during the coronavirus pandemic	0.47	0.28	0.132	Moderate	10	62
Extc 11	Accumulating interest on loans to the contractor due to work interruption during the coronavirus pandemic	0.54	0.33	0.176	Moderate	2	16

Risk Code	Risk Factor	P	I	RS	Risk Case	(RRC)	(RRT)
	External risk [others]			TRS=0.145			
Ex 1	The difficulty of expropriating land opposing the path of the roads	0.49	0.39	0.193	High	3	8
Ex 2	Difficulty and numerous procedures in case of overlap between the road paths with other parties (Antiquities-Electricity-Roads-Water)	0.56	0.43	0.238	High	1	1
Ex 3	Numerous industrial obstacles (military installations - mosques - houses -etc.) in the path of the road	0.34	0.28	0.097	Moderate	19	114
Ex 4	The remote location of the site	0.43	0.37	0.160	Moderate	10	30
Ex 5	The interferences of a pipeline, electrical and instrumentation cables with the road paths	0.56	0.39	0.216	High	2	4
Ex 6	The difficulty in obtaining work permits	0.47	0.35	0.166	Moderate	8	23
Ex 7	The absence of a map of the facilities in the project's path makes it difficult to locate them with high accuracy	0.53	0.34	0.178	Moderate	5	14
Ex 8	The climate effects such as wind, rain	0.36	0.27	0.098	Moderate	17	110
Ex 9	Unavailability of utilities on-site (such as water, electricity, telephone, etc.)	0.44	0.23	0.102	Moderate	16	104
Ex 10	The changes in legislation and the laws of the government	0.35	0.27	0.094	Moderate	20	119
Ex 11	Fluctuations in the exchange rate	0.44	0.27	0.122	Moderate	15	78
Ex 12	Float Egyptian currency	0.56	0.33	0.185	High	4	11
Ex 13	Difficulty obtaining foreign currency from banks to buy supplies that are not available inside Egypt	0.51	0.33	0.169	Moderate	7	22
Ex 14	The change in financial policy for loaning	0.49	0.30	0.147	Moderate	12	43
Ex 15	Increasing in taxes	0.53	0.29	0.156	Moderate	11	33
Ex 16	Revolution, strikes, war, and closed regions	0.32	0.31	0.097	Moderate	18	112
Ex 17	Natural disasters (torrents, earthquakes, etc.)	0.25	0.32	0.078	Moderate	22	144
Ex 18	Accidents during implementation	0.33	0.25	0.083	Moderate	21	137
Ex 19	Difficulty in implementation due to traffic (maintenance projects)	0.47	0.30	0.140	Moderate	13	53
Ex 20	The inflation rate and its effects on the raising of prices	0.53	0.31	0.165	Moderate	9	24

Risk Code	Risk Factor	P	I	RS	Risk Case	(RRC)	(RRT)
Ex 21	Difficulties in obtaining construction materials at official current prices	0.47	0.27	0.127	Moderate	14	70
Ex 22	The delay in receiving the materials from suppliers	0.44	0.39	0.170	Moderate	6	21
	Operational Risk	TRS=0.187					
Op 1	Excess and illegal loads on the implemented roads	0.63	0.35	0.219	High	1	3
Op 2	Weakness of linking the project to accident records	0.53	0.29	0.155	Moderate	2	34

It can be said that, the high-risk factors that influence project delays are the following:

- Difficulty and numerous procedures in case of overlap between the road paths with other parties (Antiquities-Electricity-Roads-Water) [RR=1]
- Delayed and stopped the supplies inside and outside Egypt due to the closure of the countries [RR=2]
- Excess and illegal loads on the implemented roads [RR=3]
- The interferences of a pipeline, electrical and instrumentation cables with the road paths [RR=4]
- Difficulties in the financial funding of the contractor [RR=5]
- Using old equipment with low rates, with no attention to periodic maintenance [RR=6]
- The design doesn't finish at the beginning of the project [RR=7]
- The difficulty of expropriating land opposing the path of the roads [RR=8]
- Delays in fixing the equipment [RR=9]
- Equipment shortage [RR=10]
- Float Egyptian currency [RR=11]
- Tender award for the lowest price regardless of the contractor's competence to complete the project on time and with the required quality [RR=12]
- Frequent equipment failures and lack of spare parts [RR=13]

8. DEMATEL Method

After evaluating the risk score [RS] for the factors, it is impossible to mitigate all the risk factors at once with limited funds. As shown in Table 5, five significant factors were identified for each risk category to more effectively mitigate the impact of risk factors. In addition, the factors with the highest score degrees from all risks have been identified for further investigation, as shown in Table 6. As a result, the second questionnaire was designed for use with the DEMATEL (decision-making trial and evaluation laboratory) method. It was also released in order to establish the causal relationships between the risk factors.

Table 5. the high-risk factors in each category

RRC	Risk category (RC)	Risk category (RC)	Risk category (RC)
	Consultant	Design	Equipment
1	Con 4	Des 11	Eq 12
2	Con 5	Des 2	Eq 3
3	Con1	Des 13	Eq 5
4	Con 2	Des 14	Eq 1
5	Con 11	Des 3	Eq 8

RRC	Risk category (RC) Material	Risk category (RC) Owner	Risk category (RC) Contract
1	Ma 6	Ow 4	Contr 4
2	Ma 9	Ow 9	Contr 3
3	Ma 10	Ow 12	Contr 7
4	Ma 1	Ow 2	Contr 5
5	Ma 7	Ow 11	Contr 6
	Contractor	Labor	External [coronavirus effects]
1	Cont1	Lab 2	Extc 5
2	Cont 22	Lab 1	Extc 11
3	Cont 7	Lab 8	Extc 3
4	Cont 24	Lab 9	Extc 2
5	Cont 16	Lab 10	Extc 7
	External [others]	Operational	
1	Ex 2	Op 1	
2	Ex 5	Op 2	
3	Ex 1		
4	Ex 12		
5	Ex 7		

Table 6. High-risk factors for all identified factors [High risk score]

RRT	**RC**	**RRT**	**RC**
1	**Ex 2**	8	**Ex 1**
2	**Extc 5**	9	**Eq 3**
3	**Op 1**	10	**Eq 5**
4	**Ex 5**	11	**Ex 12**
5	**Cont1**	12	**Ow 4**
6	**Eq 12**	13	**Eq 1**
7	**Des 11**		

Tzeng [27] and Wu [28] describe the DEMATEL method's procedure as follows:
First step: Create an average matrix. And, nineteen experts' judgments serve as the basis for the computation utilizing the DEMATEL approach. Each respondent was asked to rate the direct influence of any two factors using an integer score between 0 and 3, representing "no effect," "low effect," "medium effect," and "high effect," respectively. To include all the viewpoints from R respondents, the average matrix M_{avg} = [Mij] can be built as follows:

$$Mij = \frac{1}{R}\sum_{k=1}^{R} x_{ij}^{k} \qquad (3)$$

Where, k is the number of responders from 1 to R.
Based on Equation 3, the average matrix M_{avg} for the first five high-risk in each category can be calculated as following:

IOP Conf. Series: Materials Science and Engineering 1269 (2022) 012007 doi:10.1088/1757-899X/1269/1/012007

Consultant

	Con4	Con5	Con1	Con2	Con11
Con4	0.0000	2.7895	2.8421	2.6842	2.6316
Con5	0.0000	0.0000	2.6842	2.4737	2.3684
Con1	0.1053	2.0000	0.0000	1.7368	1.8947
Con2	0.0000	1.0526	2.4211	0.0000	2.5789
Con11	0.0000	0.0526	2.6316	2.4211	0.0000

Design

	Des11	Des2	Des13	Des14	Des3
Des 11	0.0000	2.3684	0.4211	0.2105	0.0526
Des 2	2.3158	0.0000	0.7895	0.0526	0.0000
Des 13	0.2632	1.4211	0.0000	0.2632	0.0526
Des 14	0.0000	0.3684	0.0000	0.0000	0.0000
Des 3	2.7368	2.2632	2.4211	1.5789	0.0000

Equipment

	Eq12	Eq3	Eq5	Eq1	Eq8
Eq12	0.0000	2.2105	2.2632	2.8421	0.6316
Eq3	0.8947	0.0000	2.0526	1.7368	0.2632
Eq5	0.2632	0.1579	0.0000	0.2105	0.3158
Eq1	0.5263	1.3158	2.0000	0.0000	0.1579
Eq8	0.5263	0.6316	1.8947	0.6316	0.0000

Material

	M 6	Ma9	Ma10	Ma1	Ma7
Ma 6	0.0000	2.2105	2.1579	2.7368	2.0000
Ma 9	1.2105	0.0000	1.1053	2.6316	1.9474
Ma 10	1.2105	0.5263	0.0000	2.3158	2.0000
Ma 1	0.5263	0.7368	0.3158	0.0000	1.6842
Ma 7	1.1579	0.7895	1.4211	1.7895	0.0000

Owner

	Ow 4	Ow 9	Ow 12	Ow 2	Ow 11
Ow4	0.0000	0.0000	0.0000	0.0000	0.0000
Ow9	0.0000	0.0000	2.5263	1.0526	2.6842
Ow12	0.0000	2.2632	0.0000	1.2105	2.6842
Ow2	0.0000	0.0000	0.0000	0.0000	0.0000
Ow11	0.0000	2.5263	2.4737	1.0526	0.0000

Contract

	Contr4	Contr3	Contr7	Contr5	Contr6
Contr4	0.0000	1.3684	0.4737	0.0000	1.8421
Contr3	2.9474	0.0000	0.6316	0.0000	2.1579
Contr7	0.1053	0.1053	0.0000	0.0000	0.0526
Contr5	2.5789	1.4211	0.5789	0.0000	1.5789
Contr6	2.6316	1.4211	0.6316	0.0000	0.0000

Contractor

	Cont1	Cont22	Cont7	Cont24	Cont16
Cont1	0.0000	2.7368	1.8421	2.5789	2.6316
Cont22	0.0000	0.0000	0.0526	2.5789	0.2105
Cont7	0.0000	1.6842	0.0000	1.6316	0.7895
Cont24	0.0000	2.4211	0.0000	0.0000	0.2105
Cont16	0.0000	1.6316	0.1053	1.4737	0.0000

Labor

	Lab2	Lab1	Lab8	Lab9	Lab10
Lab2	0.0000	2.1053	0.5263	1.2105	2.0000
Lab1	0.5789	0.0000	0.6316	1.4737	2.4211
Lab8	2.6842	1.7368	0.0000	2.4211	2.5789
Lab9	2.6316	1.2105	0.9474	0.0000	2.4211
Lab10	0.4737	1.5263	0.3684	1.2105	0.0000

External [coronavirus effects]

	Extc5	Extc11	Extc3	Extc2	Extc7
Extc5	0.0000	2.3158	1.3158	2.4737	0.5263
Extc11	2.5263	0.0000	2.3684	2.7368	0.3684
Extc3	0.5263	0.6316	0.0000	1.5263	1.0000
Extc2	2.6842	2.8421	2.3684	0.0000	1.2632
Extc7	0.8947	0.9474	1.7368	1.6842	0.0000

External [others]

	Ex2	Ex5	Ex1	Ex12	Ex7
Ex2	0.0000	0.1579	2.5263	0.0000	0.0000
Ex5	2.6316	0.0000	2.3684	0.0000	0.0526
Ex1	0.0526	0.1053	0.0000	0.0000	0.0000
Ex12	0.0000	0.0000	0.0000	0.0000	0.0000
Ex7	2.7368	1.7368	2.1053	0.0000	0.0000

Operational

	Op1	Op2
Op1	0.0000	0.0000
Op2	0.0000	0.0000

Also, the average matrix M_{avg} for high-risk ranking for all identified risk can be shown as:

	Ex2	Extc5	Op1	Ex5	Cont1	Eq12	Des11	Ex1	Eq3	Eq5	Ex12	Ow4	Eq1
Ex2	0.0000	1.7368	0.0000	0.1579	0.0000	0.0000	2.1579	2.5263	0.0000	0.0000	0.0000	0.0000	0.0000
Extc5	0.0000	0.0000	0.0000	0.0000	0.3158	0.0000	0.0000	0.0000	0.0526	0.0000	0.0000	0.0000	0.0000
Op1	0.0000	0.4737	0.0000	0.0000	0.0000	0.0000	0.0000	0.0000	0.0000	0.0000	0.0000	0.0000	0.0000
Ex5	2.6316	1.3158	0.0000	0.0000	0.2105	0.0000	2.3158	2.3684	0.0000	0.0000	0.0000	0.0000	0.0000
Cont1	0.5789	2.5263	0.0000	0.0000	0.0000	1.9474	0.0000	0.1579	2.4211	2.4737	0.0000	0.2632	2.2105
Eq 12	0.0000	0.5789	0.0000	0.0000	0.0000	0.0000	0.0000	0.2105	2.2105	2.2632	0.0000	0.0000	2.8421
Des11	1.3684	1.4211	0.0000	1.4211	0.1053	0.0000	0.0000	2.1053	0.1053	0.0000	0.0000	0.0000	0.0000
Ex1	0.0526	1.4211	0.0000	0.1053	0.1053	0.0000	1.3158	0.0000	0.0000	0.0000	0.0000	0.0000	0.0000
Eq3	0.0000	0.7895	0.0000	0.0000	0.0000	0.8947	0.0000	0.1579	0.0000	2.0526	0.0000	0.0000	1.7368
Eq5	0.0000	1.5789	0.0000	0.0000	0.0526	0.2632	0.0000	0.4211	0.1579	0.0000	0.0000	0.0000	0.2105
Ex12	0.0000	2.8421	0.1053	0.0000	2.2632	1.1053	0.1053	0.0000	1.5789	1.6316	0.0000	0.0000	0.7368
Ow4	0.2105	2.2632	0.0000	0.0000	1.2105	2.0000	0.0526	0.2105	2.0000	2.2632	0.0000	0.0000	2.1579
Eq1	0.0000	1.5263	0.0000	0.0000	0.0526	0.5263	0.0000	0.3158	1.3158	2.0000	0.0000	0.0000	0.0000

Second step: the normalized direct relation matrix Nd is calculated from the average matrix Mavg. As in the DEMATEL method, this is accomplished by dividing each element of the direct relation matrix by the greatest number of row sums. The normalized direct relation matrix [ND] can be calculated as follow:

$$N_D = \frac{1}{max_{1\le i\le n} \sum_i^n M_{ij}} \left[M_{avg}\right] \qquad (4)$$

The normalized initial direct-relation matrix D can be shown as follows:

Consultant

	Con4	Con5	Con1	Con2	Con11
Con4	0.0000	0.2548	0.2596	0.2452	0.2404
Con5	0.0000	0.0000	0.2452	0.2260	0.2163
Con1	0.0096	0.1827	0.0000	0.1586	0.1731
Con2	0.0000	0.0962	0.2212	0.0000	0.2356
Con11	0.0000	0.0048	0.2404	0.2212	0.0000

Design

	Des11	Des2	Des13	Des14	Des3
Des 11	0.0000	0.2632	0.0468	0.0234	0.0058
Des 2	0.2573	0.0000	0.0877	0.0058	0.0000
Des 13	0.0292	0.1579	0.0000	0.0292	0.0058
Des 14	0.0000	0.0409	0.0000	0.0000	0.0000
Des 3	0.3041	0.2515	0.2690	0.1754	0.0000

Equipment

	Eq12	Eq3	Eq5	Eq1	Eq8
Eq12	0.0000	0.2781	0.2848	0.3576	0.0795
Eq3	0.1126	0.0000	0.2583	0.2185	0.0331
Eq5	0.0331	0.0199	0.0000	0.0265	0.0397
Eq1	0.0662	0.1656	0.2517	0.0000	0.0199
Eq8	0.0662	0.0795	0.2384	0.0795	0.0000

Material

	M 6	Ma9	Ma10	Ma1	Ma7
Ma 6	0.0000	0.2428	0.2370	0.3006	0.2197
Ma 9	0.1329	0.0000	0.1214	0.2890	0.2139
Ma 10	0.1329	0.0578	0.0000	0.2543	0.2197
Ma 1	0.0578	0.0809	0.0347	0.0000	0.1850
Ma 7	0.1272	0.0867	0.1561	0.1965	0.0000

Owner

	Ow 4	Ow 9	Ow 12	Ow 2	Ow 11
Ow4	0.0000	0.0000	0.0000	0.0000	0.0000
Ow9	0.0000	0.0000	0.4034	0.1681	0.4286
Ow12	0.0000	0.3614	0.0000	0.1933	0.4286
Ow2	0.0000	0.0000	0.0000	0.0000	0.0000
Ow11	0.0000	0.4034	0.3950	0.1681	0.0000

Contract

	Contr4	Contr3	Contr7	Contr5	Contr6
Contr4	0.0000	0.2222	0.0769	0.0000	0.2991
Contr3	0.4786	0.0000	0.1026	0.0000	0.3504
Contr7	0.0171	0.0171	0.0000	0.0000	0.0085
Contr5	0.4188	0.2308	0.0940	0.0000	0.2564
Contr6	0.4274	0.2308	0.1026	0.0000	0.0000

Contractor

	Cont1	Cont22	Cont7	Cont24	Cont16
Cont1	0.0000	0.2796	0.1882	0.2634	0.2688
Cont22	0.0000	0.0000	0.0054	0.2634	0.0215
Cont7	0.0000	0.1720	0.0000	0.1667	0.0806
Cont24	0.0000	0.2473	0.0000	0.0000	0.0215
Cont16	0.0000	0.1667	0.0108	0.1505	0.0000

Labor

	Lab2	Lab1	Lab8	Lab9	Lab10
Lab2	0.0000	0.2235	0.0559	0.1285	0.2123
Lab1	0.0614	0.0000	0.0670	0.1564	0.2570
Lab8	0.2849	0.1844	0.0000	0.2570	0.2737
Lab9	0.2793	0.1285	0.1006	0.0000	0.2570
Lab10	0.0503	0.1620	0.0391	0.1285	0.0000

External [coronavirus effects]

	Extc5	Extc11	Extc3	Extc2	Extc7
Extc5	0.0000	0.2529	0.1437	0.2701	0.0575
Extc11	0.2759	0.0000	0.2586	0.2989	0.0402
Extc3	0.0575	0.0690	0.0000	0.1667	0.1092
Extc2	0.2931	0.3103	0.2586	0.0000	0.1379
Extc7	0.0977	0.1034	0.1897	0.1839	0.0000

External [others]

	Ex2	Ex5	Ex1	Ex12	Ex7
Ex2	0.0000	0.0240	0.3840	0.0000	0.0000
Ex5	0.4000	0.0000	0.3600	0.0000	0.0080
Ex1	0.0080	0.0160	0.0000	0.0000	0.0000
Ex12	0.0000	0.0000	0.0000	0.0000	0.0000
Ex7	0.4160	0.2640	0.3200	0.0000	0.0000

Operational

	Op1	Op2
Op1	0.0000	0.0000
Op2	0.0000	0.0000

High-risk factors

	Ex2	Extc5	Op1	Ex5	Cont1	Eq12	Des11	Ex1	Eq3	Eq5	Ex12	Ow4	Eq1
Ex2	0.0000	0.1381	0.0000	0.0126	0.0000	0.0000	0.1715	0.2008	0.0000	0.0000	0.0000	0.0000	0.0000
Extc5	0.0000	0.0000	0.0000	0.0000	0.0251	0.0000	0.0000	0.0000	0.0042	0.0000	0.0000	0.0000	0.0000
Op1	0.0000	0.0377	0.0000	0.0000	0.0000	0.0000	0.0000	0.0000	0.0000	0.0000	0.0000	0.0000	0.0000
Ex5	0.2092	0.1046	0.0000	0.0000	0.0167	0.0000	0.1841	0.1883	0.0000	0.0000	0.0000	0.0000	0.0000
Cont1	0.0460	0.2008	0.0000	0.0000	0.0000	0.1548	0.0000	0.0126	0.1925	0.1967	0.0000	0.0209	0.1757
Eq 12	0.0000	0.0460	0.0000	0.0000	0.0000	0.0000	0.0000	0.0167	0.1757	0.1799	0.0000	0.0000	0.2259
Des11	0.1088	0.1130	0.0000	0.1130	0.0084	0.0000	0.0000	0.1674	0.0084	0.0000	0.0000	0.0000	0.0000
Ex1	0.0042	0.1130	0.0000	0.0084	0.0084	0.0000	0.1046	0.0000	0.0000	0.0000	0.0000	0.0000	0.0000
Eq3	0.0000	0.0628	0.0000	0.0000	0.0000	0.0711	0.0000	0.0126	0.0000	0.1632	0.0000	0.0000	0.1381
Eq5	0.0000	0.1255	0.0000	0.0000	0.0042	0.0209	0.0000	0.0335	0.0126	0.0000	0.0000	0.0000	0.0167
Ex12	0.0000	0.2259	0.0084	0.0000	0.1799	0.0879	0.0084	0.0000	0.1255	0.1297	0.0000	0.0000	0.0586
Ow4	0.0167	0.1799	0.0000	0.0000	0.0962	0.1590	0.0042	0.0167	0.1590	0.1799	0.0000	0.0000	0.1715
Eq1	0.0000	0.1213	0.0000	0.0000	0.0042	0.0418	0.0000	0.0251	0.1046	0.1590	0.0000	0.0000	0.0000

Third step: the total relation matrix T has been measured by following relationship:

$$T = N_D(I - N_d)^{-1} \quad (5)$$

Where: [I] is the identity matrix.
The total relation matrix T can be shown as follows:

Consultant

	Con4	Con5	Con1	Con2	Con11
Con4	0.0061	0.4297	0.6323	0.5721	0.5790
Con5	0.0046	0.1313	0.4761	0.4271	0.4289
Con1	0.0118	0.2605	0.2247	0.3334	0.3497
Con2	0.0039	0.1924	0.4078	0.1965	0.3950
Con11	0.0037	0.1106	0.3869	0.3468	0.1735

Design

	Des11	Des2	Des13	Des14	Des3
Des 11	0.0816	0.3000	0.0787	0.0305	0.0068
Des 2	0.2853	0.0937	0.1099	0.0167	0.0023
Des 13	0.0795	0.1853	0.0217	0.0340	0.0064
Des 14	0.0117	0.0448	0.0045	0.0007	0.0001
Des 3	0.4241	0.4240	0.3272	0.1982	0.0044

Equipment

	Eq12	Eq3	Eq5	Eq1	Eq8
Eq12	0.1092	0.4160	0.5849	0.5138	0.1354
Eq3	0.1658	0.1135	0.4329	0.3199	0.0736
Eq5	0.0473	0.0477	0.0522	0.0591	0.0483
Eq1	0.1149	0.2268	0.3822	0.1051	0.0538
Eq8	0.1070	0.1454	0.3544	0.1614	0.0306

Material

	M 6	Ma9	Ma10	Ma1	Ma7
Ma 6	0.2404	0.4404	0.4676	0.7382	0.6059
Ma 9	0.2977	0.1831	0.3134	0.6094	0.5000
Ma 10	0.2775	0.2201	0.1843	0.5402	0.4681
Ma 1	0.1529	0.1707	0.1491	0.1970	0.3243
Ma 7	0.2569	0.2265	0.3008	0.4663	0.2572

Owner

	Ow 4	Ow 9	Ow 12	Ow 2	Ow 11
Ow4	0.0000	0.0000	0.0000	0.0000	0.0000
Ow9	0.0000	1.1792	1.5022	0.9217	1.5777
Ow12	0.0000	1.4014	1.1698	0.9121	1.5305
Ow2	0.0000	0.0000	0.0000	0.0000	0.0000
Ow11	0.0000	1.4325	1.4629	0.9001	1.2409

Contract

	Contr4	Contr3	Contr7	Contr5	Contr6
Contr4	0.4973	0.4792	0.2277	0.0000	0.6178
Contr3	1.0313	0.4212	0.3081	0.0000	0.8092
Contr7	0.0508	0.0371	0.0115	0.0000	0.0368
Contr5	1.0963	0.6697	0.3313	0.0000	0.8219
Contr6	0.8831	0.5366	0.2722	0.0000	0.4545

Contractor

	Cont1	Cont22	Cont7	Cont24	Cont16
Cont1	0.0000	0.4795	0.1940	0.4680	0.3048
Cont22	0.0000	0.0776	0.0061	0.2894	0.0299
Cont7	0.0000	0.2486	0.0023	0.2463	0.0915
Cont24	0.0000	0.2713	0.0018	0.0762	0.0291
Cont16	0.0000	0.2231	0.0121	0.2129	0.0103

Labor

	Lab2	Lab1	Lab8	Lab9	Lab10
Lab2	0.1767	0.4061	0.1429	0.3123	0.4736
Lab1	0.2254	0.1937	0.1427	0.3133	0.4742
Lab8	0.5424	0.5142	0.1458	0.5343	0.6983
Lab9	0.4573	0.3916	0.1994	0.2448	0.5722
Lab10	0.1757	0.2842	0.1007	0.2473	0.2015

External [coronavirus effects]

	Extc5	Extc11	Extc3	Extc2	Extc7
Extc5	0.4777	0.6847	0.6414	0.7636	0.2878
Extc11	0.7484	0.5386	0.7834	0.8491	0.3076
Extc3	0.3180	0.3310	0.2901	0.4428	0.2335
Extc2	0.8070	0.8245	0.8396	0.6784	0.4027
Extc7	0.4305	0.4405	0.5428	0.5551	0.1783

External [others]

	Ex2	Ex5	Ex1	Ex12	Ex7
Ex2	0.0157	0.0309	0.4012	0.0000	0.0002
Ex5	0.4159	0.0206	0.5297	0.0000	0.0082
Ex1	0.0148	0.0166	0.0117	0.0000	0.0001
Ex12	0.0000	0.0000	0.0000	0.0000	0.0000
Ex7	0.5370	0.2876	0.6305	0.0000	0.0023

Operational

	Op1	Op2
Op1	0.0000	0.0000
Op2	0.0000	0.0000

High-risk factors

	Ex2	Extc5	Op1	Ex5	Cont1	Eq12	Des11	Ex1	Eq3	Eq5	Ex12	Ow4	Eq1
Ex2	0.0325	0.2019	0.0000	0.0388	0.0096	0.0021	0.2105	0.2503	0.0052	0.0036	0.0000	0.0002	0.0030
Extc5	0.0012	0.0086	0.0000	0.0001	0.0255	0.0053	0.0004	0.0014	0.0110	0.0090	0.0000	0.0005	0.0074
Op1	0.0000	0.0380	0.0000	0.0000	0.0010	0.0002	0.0000	0.0001	0.0004	0.0003	0.0000	0.0000	0.0003
Ex5	0.2477	0.2137	0.0000	0.0353	0.0274	0.0056	0.2634	0.2899	0.0104	0.0094	0.0000	0.0006	0.0078
Cont1	0.0493	0.3208	0.0000	0.0027	0.0132	0.1972	0.0143	0.0501	0.2667	0.3247	0.0000	0.0212	0.2685
Eq 12	0.0010	0.1329	0.0000	0.0008	0.0059	0.0330	0.0042	0.0367	0.2147	0.2648	0.0000	0.0001	0.2685
Des11	0.1446	0.1870	0.0000	0.1249	0.0178	0.0043	0.0722	0.2329	0.0148	0.0078	0.0000	0.0004	0.0063
Ex1	0.0221	0.1388	0.0000	0.0219	0.0135	0.0028	0.1154	0.0284	0.0051	0.0046	0.0000	0.0003	0.0038
Eq3	0.0008	0.1204	0.0000	0.0006	0.0049	0.0859	0.0030	0.0266	0.0367	0.2122	0.0000	0.0001	0.1670
Eq5	0.0012	0.1397	0.0000	0.0008	0.0082	0.0252	0.0042	0.0365	0.0222	0.0141	0.0000	0.0002	0.0272
Ex12	0.0108	0.3422	0.0084	0.0018	0.1910	0.1449	0.0132	0.0247	0.2097	0.2535	0.0000	0.0040	0.1588
Ow4	0.0239	0.3130	0.0000	0.0023	0.1074	0.2124	0.0140	0.0503	0.2520	0.3252	0.0000	0.0022	0.2790
Eq1	0.0012	0.1676	0.0000	0.0008	0.0097	0.0578	0.0042	0.0363	0.1235	0.1971	0.0000	0.0002	0.0352

Fourth step: Calculate the sums of matrix T's rows and columns. And, the sum of rows and the sum of columns are represented by vectors R and C, respectively. And, if R_i is the sum of the ith row in matrix T, it describes both the direct and indirect influences of factor i on the other factors. Also, if C_j is the sum of the jth column in matrix T, it represents both direct and indirect effects of factor j on the other factors. When j = I, the value of ($R_i + C_i$) represents the total effect of factor i, both given and received. And the difference (R_i - C_j) represents the factor's net contribution to the system. Tables 7 and 8 show the values of both $R_i + C_i$ and $R_i - C_j$.

Table 7. the values of Ri+Ci and Ri-Ci [risk category]

RC	R_i+C_i	R_i-C_i	Cause or effect	RC	R_i+C_i	R_i-C_i	Cause or effect
Consultant				Design			
Con4	2.2492	2.1890	Cause	Des 11	1.3798	-0.3845	Effect
Con5	2.5925	0.3435	Cause	Des 2	1.5557	-0.5397	Effect
Con1	3.3078	-0.9477	Effect	Des 13	0.8689	-0.2152	Effect
Con2	3.0716	-0.6803	Effect	Des 14	0.3418	-0.2183	Effect
Con11	2.9476	-0.9046	Effect	Des 3	1.3978	1.3578	Cause
Equipment				Material			
Eq12	2.3035	1.2150	Cause	Ma6	3.7181	1.2671	Cause
Eq3	2.0551	0.1563	Cause	Ma9	3.1444	0.6627	Cause
Eq5	2.0610	-1.5519	Effect	Ma10	3.1052	0.2750	Cause
Eq1	2.0420	-0.2764	Effect	Ma1	3.5450	-1.5570	Effect
Eq8	1.1405	0.4571	Cause	Ma7	3.6631	-0.6478	Effect
Owner				Contract			
Ow 4	0.0000	0.0000	Cause	Contr4	5.3807	-1.7368	Effect
Ow 9	9.1940	1.1677	Cause	Contr3	4.7136	0.4260	Cause
Ow 12	9.1487	0.8790	Cause	Contr7	1.2870	-1.0146	Effect
Ow 2	2.7340	-2.7340	Effect	Contr5	2.9192	2.9192	Cause
Ow 11	9.3856	0.6873	Cause	Contr6	4.8866	-0.5938	Effect

RC	R_i+C_i	R_i-C_i	Cause or effect	RC	R_i+C_i	R_i-C_i	Cause or effect
Contractor				Labor			
Cont1	1.4464	1.4464	Cause	Lab 2	3.0891	-0.0659	Effect
Cont 22	1.7031	-0.8971	Effect	Lab 1	3.1392	-0.4407	Effect
Cont 7	0.8050	0.3724	Cause	Lab 8	3.1665	1.7036	Cause
Cont 24	1.6711	-0.9144	Effect	Lab 9	3.5172	0.2133	Cause
Cont 16	0.9241	-0.0073	Effect	Lab 10	3.4290	-1.4103	Effect
External [coronavirus effects]				External [others]			
Extc 5	5.6368	0.0736	Cause	Ex 2	1.4313	-0.5354	Effect
Extc 11	6.0462	0.4078	Cause	Ex 5	1.3300	0.6187	Cause
Extc 3	4.7128	-1.4819	Effect	Ex 1	1.6163	-1.5299	Effect
Extc 2	6.8412	0.2633	Cause	Ex 12	0.0000	0.0000	Cause
Extc 7	3.5571	0.7372	Cause	Ex 7	1.4682	1.4466	Cause
Operational							
Op 1	0.0000	0.0000	Cause				
Op 2	0.0000	0.0000	Cause				

Table 8. the values of Ri+Ci and Ri-Ci [High-risk factors]

RC	Ri+Ci	Ri-Ci	Identity
Ex 2	1.2941	0.2214	Cause
Extc 5	2.3948	-2.2540	Effect
Op 1	0.0487	0.0319	Cause
Ex 5	1.3419	0.8805	Cause
Cont1	1.9637	1.0935	Cause
Eq 12	1.7394	0.1859	Cause
Des 11	1.5315	0.0940	Cause
Ex 1	1.4209	-0.7074	Effect
Eq 3	1.8307	-0.5144	Effect
Eq 5	1.9060	-1.3469	Effect
Ex 12	1.3629	1.3629	Cause
Ow 4	1.6118	1.5517	Cause
Eq 1	1.8663	-0.5993	Effect

Fifth Step: Create a cause-and-effect relationship diagram and set a threshold value. On the other hand, a threshold value is required for a decision maker to filter out some insignificant effects. And, only effects bigger than the threshold value would be selected and displayed in the digraph in this manner. In addition, the threshold value in this study is determined by taking the average of the elements in matrix T. The diagram can be obtained by tracing the range of data of (Ri+Ci, Ri-Ci). Furthermore, the method's output tables and graphics are intended to make the intricate linkages in a process more understandable. Figure 4 show the values of Ri-Ci for all categories and also for high-risk factors.

IOP Conf. Series: Materials Science and Engineering 1269 (2022) 012007 doi:10.1088/1757-899X/1269/1/012007

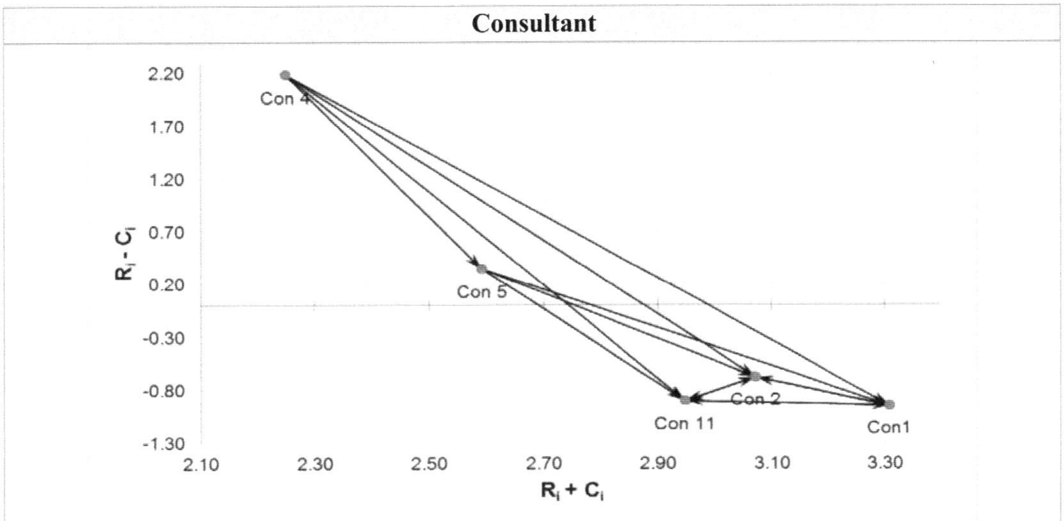

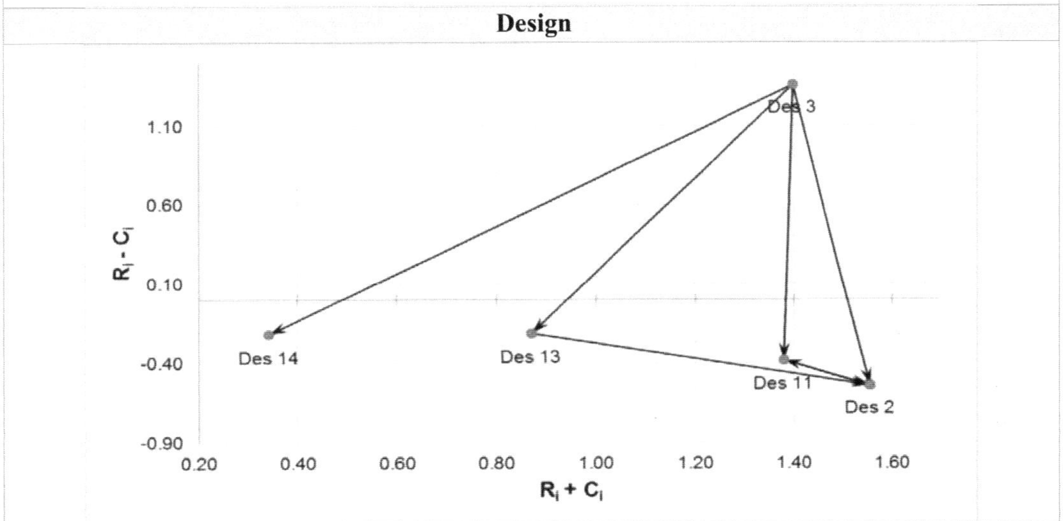

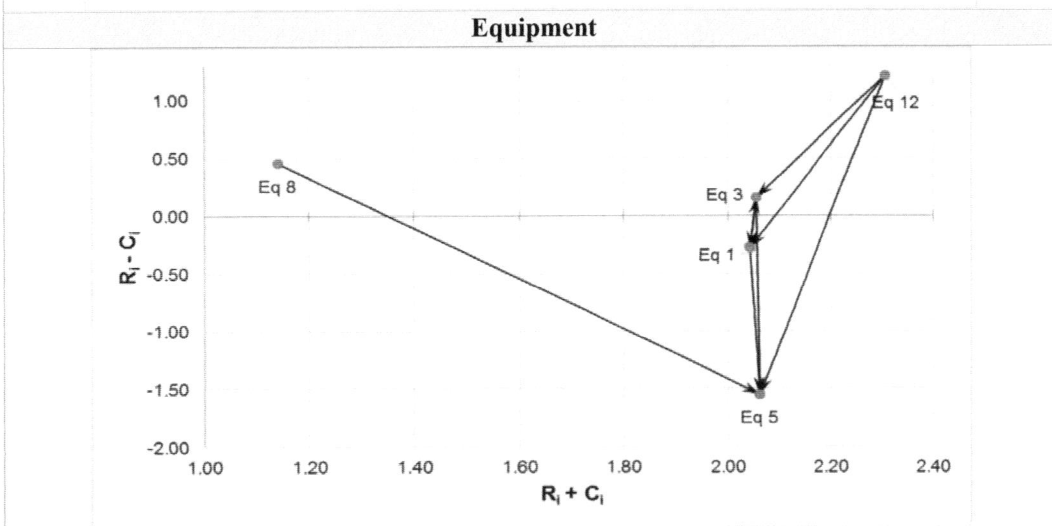

Material

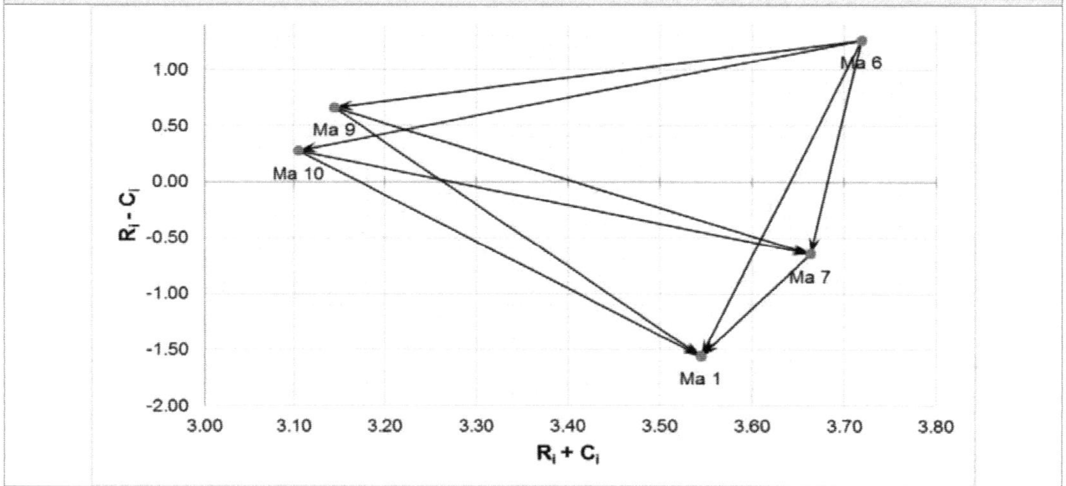

Owner

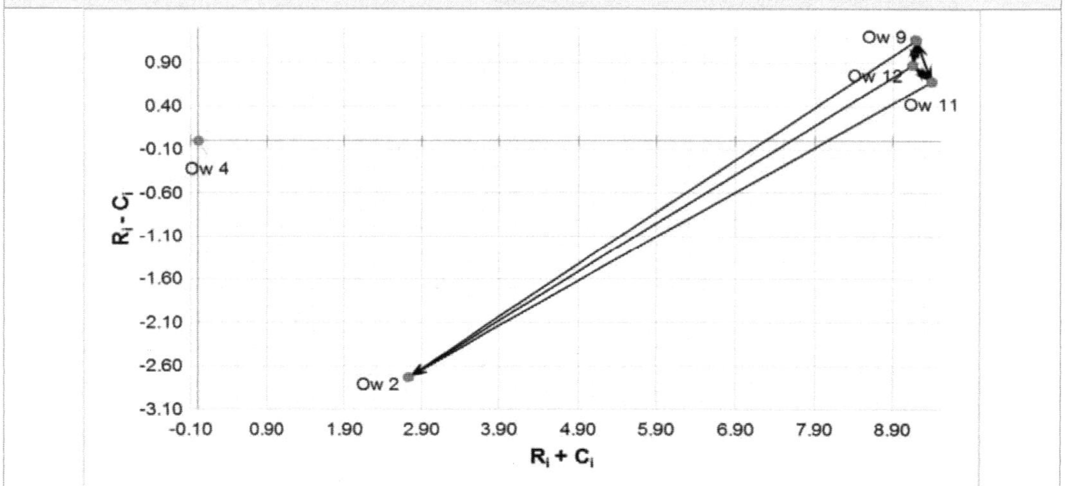

Contract

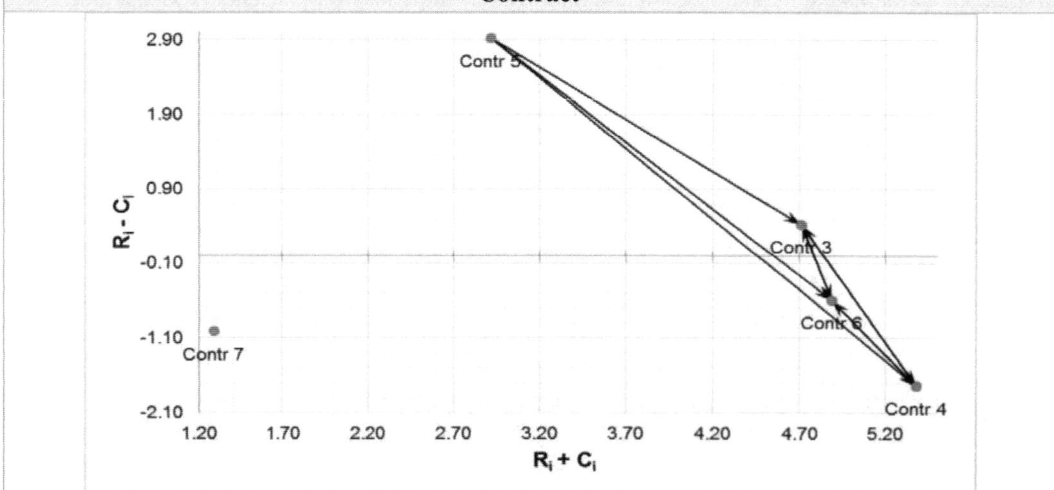

IOP Conf. Series: Materials Science and Engineering 1269 (2022) 012007 doi:10.1088/1757-899X/1269/1/012007

Contractor

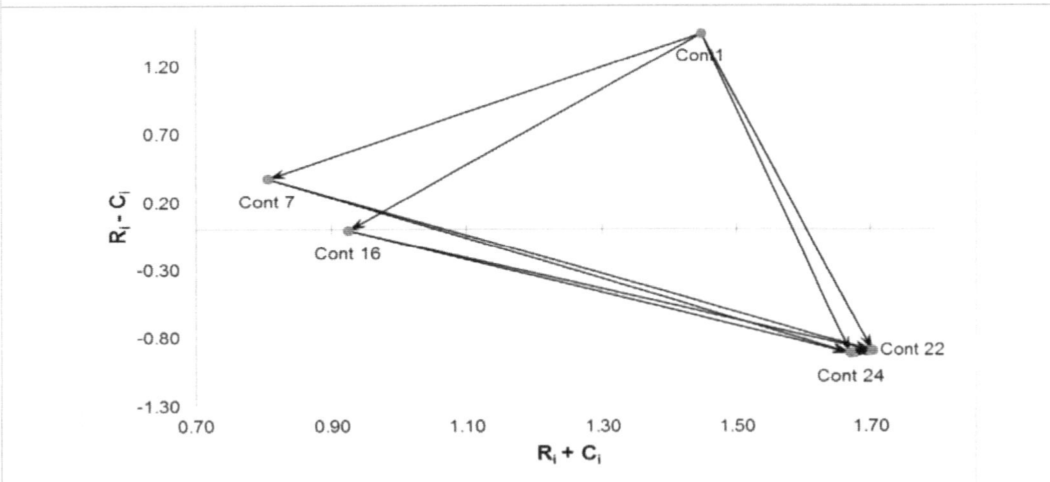

Labor

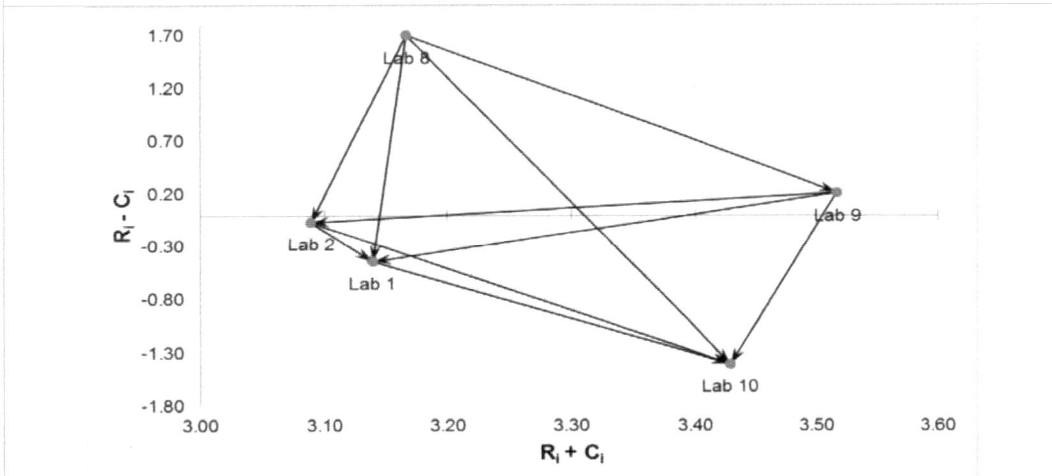

External [coronavirus effects]

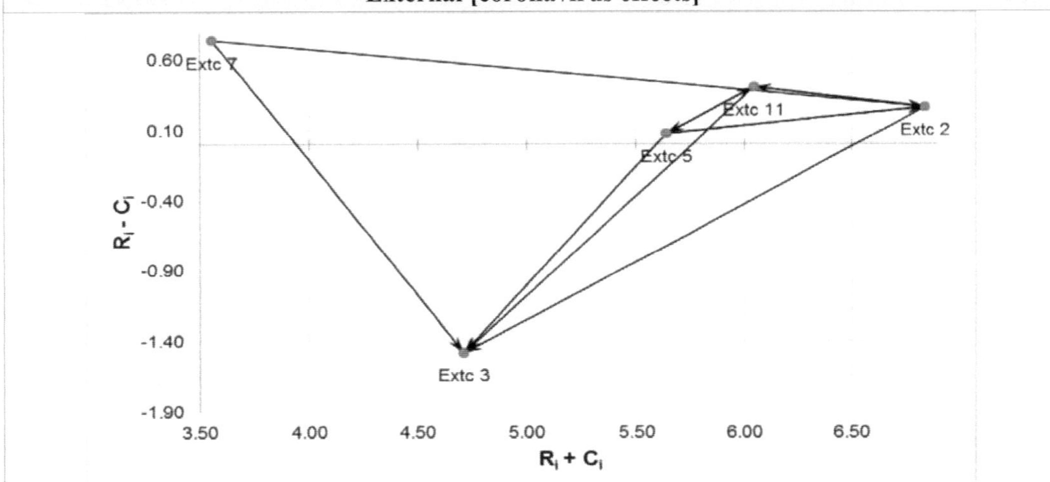

IOP Conf. Series: Materials Science and Engineering 1269 (2022) 012007 doi:10.1088/1757-899X/1269/1/012007

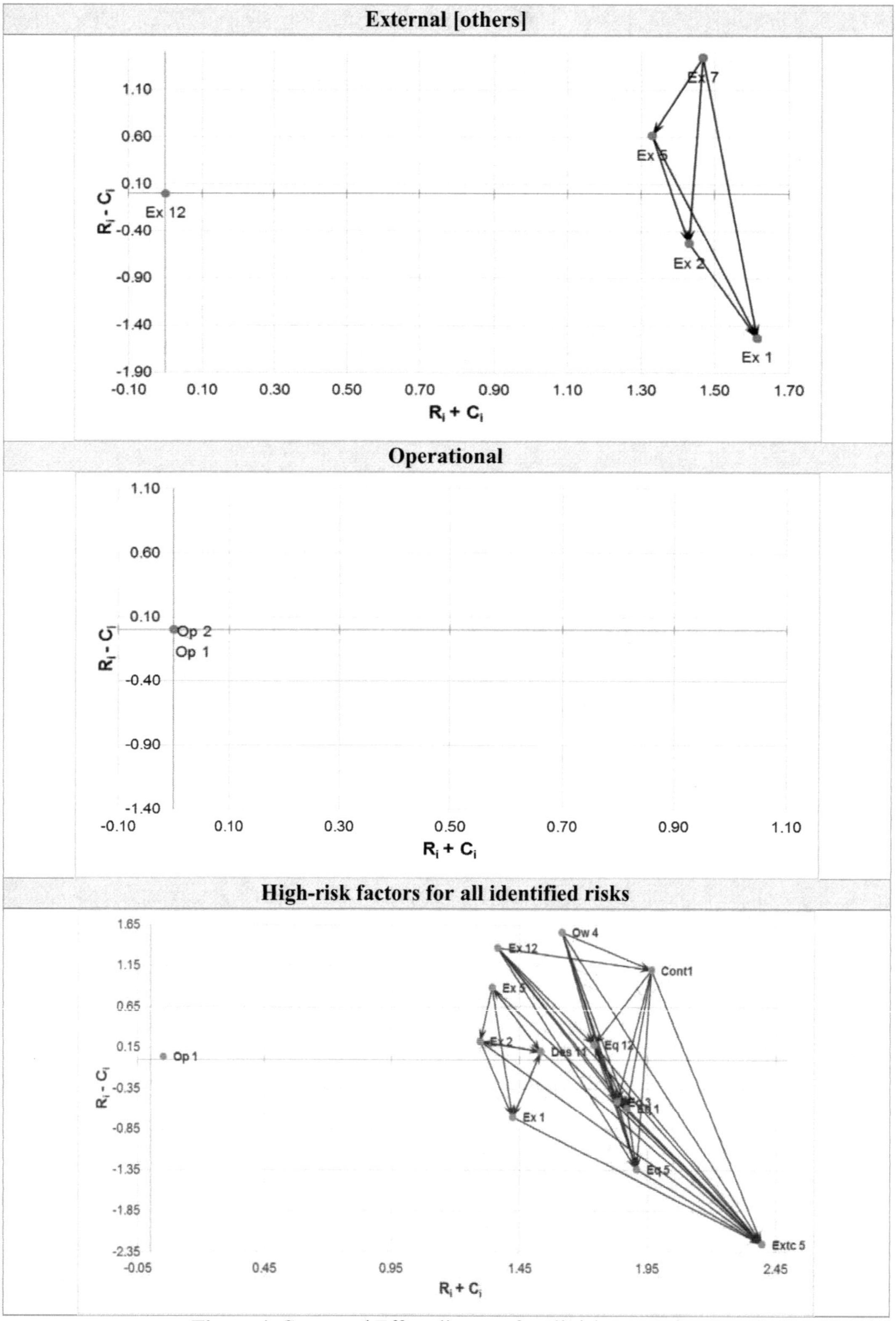

Figure 4. Cause and Effect diagram for all risk categories

The direct and indirect effects of the high-risk factors for each category (consultant, design, equipment, material, owner, contract, contractor, labor, external [coronavirus effects and others] and operational) are shown in Table 7. Additionally, Table 8 displays the outcomes of the first 13 risk factors. The average of the matrix T's elements was used to calculate the threshold values. And Table 9 displays these numbers.

Table 9. the threshold values for all categories

Ser	Type of Category	The threshold value
1	Consultant	0.2834
2	Design	0.1109
3	Equipment	0.1920
4	Material	0.3435
5	Owner	0.6092
6	Contract	0.3837
7	Contractor	0.1310
8	Labor	0.3268
9	External [coronavirus effects]	0.5359
10	External [others]	0.1169
11	Operational	0.0000
12	High-risk factors for all identified risks	0.0601

The diagram and table above show that the following criteria can be used to evaluate each risk factor:

First category [Consultant]: In terms of importance, (R_i + C_i) represents the influence of factor I on the entire system as well as the influence of other system components on the factor. For ranking, Con 1 came in first, followed by Con 2, Con 11, Con 5 and Con 4. On the other hand, the vertical vector (Ri-Ci) shows how much a factor has an impact on the system. On the whole, a causal variable is represented by a positive value of Ri+Ci, while an effect is represented by a negative value of Ri-Ci. Con 4 and Con 5 are viewed as causal variables in this study, while Con 1, Con 2, and Con 11 are seen as effects.

Second category [Design]: Ri+Ci's ranking placed Des 2 at the top, followed by Des 3, Des 11, Des 13, and Des 14. In this study Des 3 is seen as causal variables, whereas Des 11, Des 2, Des 13 and Des 14 are viewed as effects, based on the computation of (Ri-Ci).

Third category [Equipment]: Ri+Ci's ranking placed Eq 12 at the top, followed by Eq 5, Eq 3, Eq 1 and Eq 8. In this study, Eq 12, Eq 3 and Eq 8 are seen as causal variables, whereas Eq 5 and Eq 1 are viewed as effects, based on the computation of (Ri-Ci).

Fourth category [Material]: For ranking according to Ri+Ci, Ma 6 came in first, followed by Ma 7, Ma 1, Ma 9 and Ma 10. From the calculation of (Ri-Ci), Ma 6, Ma 9, and Ma 10 are viewed as causal variables in this study, while Ma 1 and Ma 7 are seen as effects.

Fifth category [Owner]: According to Ri+Ci, Ow 11 was ranked first, followed by Ow 9, Ow 12, Ow 2 and Ow 4. Based on the calculation of the (Ri-Ci), Ow 4, Ow 9, Ow 12 and Ow 11are seen as causal factors in this study, whereas Ow 2 is viewed as effects.

Sixth category [Contract]: R_i+C_i scored Contr 4 as the highest, followed by Contr 6, Contr 3, Contr 5 and Contr 7. The (Ri-Ci) calculation indicates that Contr 3 and Contr 5 are considered to be causal factors in this study, whilst Contr 4, Contr 7 and Contr 6 are considered to be effects.

Seventh category [Contractor]: According to Ri+Ci, Cont 22 was ranked first, followed by Cont 24, Cont1, Cont 16, and Cont 7. Based on the calculation of the (Ri-Ci), Cont 1 and Cont 7 are seen as

causal factors in this study, whereas Cont 22, Cont 24 and Cont 16 is viewed as effects.
Eighth category [Labor]: For ranking according to Ri+Ci, Lab 9 came in first, followed by Lab 10, Lab 8, Lab 1 and Lab 2. From the calculation of (Ri-Ci), Lab 8 and Lab 9 are viewed as causal variables in this study, while Lab 2, Lab 1 and Lab 10 are seen as effects.
Ninth category [External (coronavirus effects)]: According to Ri+Ci, Extc 2 was ranked first, followed by Extc 11, Extc 5, Extc 3 and Extc 7. Based on the calculation of the (Ri-Ci), Extc 5, Extc 11, Extc 2 and Extc 7 are seen as causal factors in this study, whereas Extc 3 is viewed as effects.
Tenth category [External risks (others)]: Ri+Ci's ranking placed Ex 1 at the top, followed by , Ex 7, Ex 2, Ex 5 and Ex 12. In this study, Ex 5, Ex 12 and Ex 7 are seen as causal variables, whereas Ex 2 and Ex 1 are viewed as effects, based on the computation of (Ri-Ci).
Eleventh category [Operational]: In this study, Op1 and Op2 are considered to be causal variables, according to the computation of the (Ri-Ci).
High-risk factors for all identified factors: For ranking according to Ri+Ci, Extc 5 came in first, followed by Cont 1, Eq 5, Eq 1, Eq 3, Eq 12, Ow 4, Des 11, Ex 1, Ex 12, Ex 5, Ex 2 and Op 1. From the calculation of (Ri-Ci), Ex 2, Op 1, Ex 5, Cont1, Eq 12, Des 11, Ex 12 and Ow 4 are viewed as causal variables in this study, while Extc 5, Ex 1, Eq 3, Eq 5 and Eq 1 are seen as effects.

In general, the strategy employed in the paper is to incorporate the DEMATEL technique and risk analysis into the statistical model. Whereas the risk analysis prioritizes risk based on the evaluation of each specific project risk. Additionally, this approach has been able to create connections between risks, which may make it easier to understand the likelihood and severity of project risk than if the risks were just considered isolated events.

9. Conclusion

Road construction is a complex process with many crucial stages during which risk factors that affect execution delay may be overlooked. Furthermore, the construction industry is confronted with numerous risk factors that can contribute to project delays, particularly during COVID-19. Furthermore, as the COVID-19 pandemic spread across the world, the impact of numerous risk factors has changed the delays in road projects. As a result, previously unknown factors [coronavirus effects] are included in the study. In addition, the study sought to investigate the new severity of risk factors associated with coronavirus infection. And, the study classified the risks associated with Egyptian road projects into eleven categories. It consists of consultant risk, design risk, equipment risk, material risk, owner risk, contract risk, contractor risk, labor risk, external risk [coronavirus effects and others] and operational risk. As well, eleven main categories are divided into 162 risk factors. To apply risk management strategies for prevention, reducing, or transferring these risks, it is essential to recognize the main risk factors that contribute to the delay in the project duration. Therefore , the risk characteristics are assessed and evaluated in the study, and risks are prioritised according to these attributes. On the whole, the findings of the investigations show that:

- The following are the high-risk factors that influence project delays:
  1. Difficulty and numerous procedures in case of overlap between the road paths with other parties (Antiquities-Electricity-Roads-Water) [RR=1]
  2. Delayed and stopping the supplies inside and outside Egypt due to the closure of the countries [RR=2]
  3. Excess and illegal loads on the implemented roads [RR=3]
  4. The interferences of a pipeline, electrical and instrumentation cables with the road paths [RR=4]
  5. Difficulties in the financial funding of the contractor [RR=5]
  6. Using old equipment with low rates, with no attention to periodic maintenance of this equipment [RR=6]
  7. The design doesn't finish at the beginning of the project [RR=7]
  8. The difficulty of expropriating land opposing the path of the roads [RR=8]
  9. Delays in fixing the equipment [RR=9]

10. Shortage in the equipment [RR=10]
11. Float Egyptian currency [RR=11]
12. Tender award for the lowest price regardless of the contractor's competence to complete the project on time and with the required quality [RR=12]
13. Frequent equipment failures of equipment and lack of spare parts [RR=13]

- The direct and indirect effects of the high-risk factors for each category have been studied. By using the DEMATEL technique, it is possible to establish the causal relationships between the risk factors in each category and determine the significance of first five factors. Also, for the first thirteen High-risk factors from all groups of risk, Ex 2, Op 1, Ex 5, Cont1, Eq 12, Des 11, Ex 12 and Ow 4 are viewed as causal variables in this study, while Extc 5, Ex 1, Eq 3, Eq 5 and Eq 1 are seen as effects.
- So, the study categorized the high-risk factors into causes and effects. In addition, risk causes rather than risk effects should be addressed. As a result, project managers should be driven to take action on risk causes rather than risk effects to save time and cost. Therefore, the statistical model was developed to incorporate the DEMATEL approach and risk analysis. Where, the risk analysis focuses on the evaluation and assessment of the attributes of each individual identified project risk and prioritizes risk. And, this method has been able to establish linkages between risks, which may help to better comprehend the likelihood and severity of project risk than if the risk were solely considered as independent occurrences.

References:

[1] Donaldson D. Railroads of the Raj: Estimating the impact of transportation infrastructure. American Economic Review, V. 108, No. 4-5. 2018, pp. 899-934.
[2] Diab A, Hafez S, Aziz R.The Use of Simulation to Predict (CFA) Equipment Productivity. Book The Use of Simulation to Predict (CFA) Equipment Productivity, Editor, ed.^eds., Master thesis, December 20th, 2004, Structural Engineering Department …, City, 2004, pp.
[3] Guide A.Project management body of knowledge (pmbok® guide). Book Project management body of knowledge (pmbok® guide), 11, Editor, ed.^eds., City, 2001, pp. 7-8.
[4] Kemshall H. Good practice in risk assessment and risk management. Jessica Kingsley Publishers; 1996.
[5] Raftery J. Risk analysis in project management. Routledge; 2003.
[6] Stumpf GR. Schedule delay analysis. Cost Engineering, V. 42, No. 7. 2000, pp. 32.
[7] Project Management I. A guide to the project management body of knowledge (PMBOK guide). 2002.
[8] Smith NJ, Merna T, Jobling P. Managing risk in construction projects. John Wiley & Sons; 2014.
[9] Ehsan N, Mirza E, Alam M, Ishaque A.Notice of Retraction: Risk management in construction industry. Book Notice of Retraction: Risk management in construction industry, 9, Editor, ed.^eds., Ieee, City, 2010, pp. 16-21.
[10] Odabaşı E.Models for estimating construction duration: An application for selected buildings on the Metu campus. Book Models for estimating construction duration: An application for selected buildings on the Metu campus, Editor, ed.^eds., Middle East Technical University, City, 2009, pp.
[11] Enshassi A, Mayer PE. Managing risks in construction projects, 18th Internationales Deutsches Projekt Management Forum. Ludwig burg, Germany. 2001.
[12] Hosny HE, Ibrahim AH, Fraig RF. Risk management framework for Continuous Flight Auger piles construction in Egypt. Alexandria engineering journal, V. 57, No. 4. 2018, pp. 2667-77.
[13] Mubarak SA. Construction project scheduling and control. John Wiley & Sons; 2015.
[14] Tang W, Qiang M, Duffield CF, Young DM, Lu Y. Risk management in the Chinese construction industry. Journal of construction engineering and management, V. 133, No. 12. 2007, pp. 944-56.

[15] Kuo Y-C, Lu S-T. Using fuzzy multiple criteria decision making approach to enhance risk assessment for metropolitan construction projects. International Journal of Project Management, V. 31, No. 4. 2013, pp. 602-14.
[16] Gamal YAS, Abd Elrazek M.Using the fault tree for analysis the schedule delay factors. Book Using the fault tree for analysis the schedule delay factors, 762, Editor, ed.^eds., IOP Publishing, City, 2020, pp. 012009.
[17] Abd Elrazek M, Gamal YAS. Study Of High Impact Factors Effecting on Delaying Construction Projects in Egypt.
[18] Triantaphyllou E. Multi-criteria decision making methods. Multi-criteria decision making methods: A comparative study, Springer, 2000, pp. 5-21.
[19] Gul M, Ak MF. Occupational Risk Assessment for Flight Schools: A 3, 4-Quasirung Fuzzy Multi-Criteria Decision Making-Based Approach. Sustainability, V. 14, No. 15. 2022, pp. 9373.
[20] Shaffiee Haghshenas S, Shaffiee Haghshenas S, Abduelrhman MA, Zare S, Mikaeil R. Identifying and Ranking of Mechanized Tunneling Project's Risks by Using A Fuzzy Multi-Criteria Decision Making Technique. Journal of Soft Computing in Civil Engineering, V. 6, No. 1. 2022, pp. 29-45.
[21] Boateng P, Chen Z, Ogunlana SO. An Analytical Network Process model for risks prioritisation in megaprojects. International journal of project management, V. 33, No. 8. 2015, pp. 1795-811.
[22] Gabus A, Fontela E. World problems, an invitation to further thought within the framework of DEMATEL. Battelle Geneva Research Center, Geneva, Switzerland. 1972, pp. 1-8.
[23] Giri BC, Molla MU, Biswas P. Pythagorean fuzzy DEMATEL method for supplier selection in sustainable supply chain management. Expert Systems with Applications, V. 193. 2022, pp. 116396.
[24] Dehghani H, Hormozi A, Nikpour M. Systematic risks assessment of precast concrete canal in irrigation projects using DEMATEL method. International Journal of System Assurance Engineering and Management, V. 13, No. 1. 2022, pp. 123-30.
[25] Zhu F, Hu H, Xu F. Risk assessment model for international construction projects considering risk interdependence using the DEMATEL method. Plos one, V. 17, No. 5. 2022, pp. e0265972.
[26] Shiri A, Hashemzadeh Khorasgani G, Akhoundi N, Jamalou F. Analysis and identification of managerial requirements for Iran's hand-made carpet industry using the Fuzzy DEMATEL method. Journal of Textile Science and Technology. 2022.
[27] Tzeng G-H, Chiang C-H, Li C-W. Evaluating intertwined effects in e-learning programs: A novel hybrid MCDM model based on factor analysis and DEMATEL. Expert systems with Applications, V. 32, No. 4. 2007, pp. 1028-44.
[28] Wu W-W. Choosing knowledge management strategies by using a combined ANP and DEMATEL approach. Expert systems with applications, V. 35, No. 3. 2008, pp. 828-35.

IOP Conf. Series: Materials Science and Engineering 1269 (2022) 012008 doi:10.1088/1757-899X/1269/1/012008

Investigation The Effect of Particle Size and Annealing on the Performance of Nano/ Micro Sized ZnO-Based Varistors

Hend Alharthi [1], Rabab Sendi[2], Zaki Mohammed[3], Talal Althagafi[4]
[1] Department of Physics, College of Science, Umm Al-Qura University, Makkah Saudi Arabia
[2] Department of Physics, College of Science, Umm Al-Qura University, Makkah Saudi Arabia
[3] Department of Chemistry, College of Science, Taif University, Taif Saudi Arabia
[4] Department of Physics, College of Science, Taif University, Taif, Saudi Arabia

Email: S44286221@st.uqu.edu.sa

Abstract. In this study, the influence of different ZnO particle sizes and the oxygen annealing process on the microstructure, and electrical properties of ZnO-based varistors was investigated. ZnO-Bi_2O_3-Mn_2O_3 discs based varistors with different size of ZnO particles were prepared by conventional ceramic method. Then half of varistors discs were annealed in oxygen atmosphere at 700 °C. SEM, Raman spectroscopy, and current-voltage (I–V) measurement were used. The results showed a significant improvement in distribution and growth of grain sizes in nano-size samples compared to that of micro-size samples. The electrical properties including nonlinear coefficient and breakdown voltage improved to 49 and 2555 V in nano-size compared with 32.5 and 201V in micro-size.
Keyword: ZnO-Based Varistor, Annealing in oxygen, Breakdown voltage, Electric properties

1. Introduction

ZnO has attracted considerable interest in recent years due to its attractive properties as semiconductors including non-toxicity, processing from solution, growth temperature for high quality ZnO (less than 700 °C), direct band-gap of ~ 3.37 eV, and large exciton binding energy of ~ 60 meV. Moreover, due to ZnO is insensitive to visible light, it is a suitable at ambient pressure and temperature. Oxide ceramic semiconductor materials based on ZnO has widely used in last view years as varistors. ZnO varistors have superior electrical characteristics, such as a voltage gradation, high nonlinear coefficient, the ability to absorb large amounts of energy, a minimal leakage current, etc [1,2]. It is usually used in various electrical circuits to protect electronic parts from the effects of sudden increases in electric current and in surge arrester [5,6,7,8]. On other hand, varistors based on ZnO suffer from some phenomena such as leakage current and consequent Joule heating which leads to degrades the ZnO varistors. To improve the electrical behaviors of the ZnO-varistor system, ZnO varistors often include one or more of metal oxides such as Bi_2O_3, Sb_2O_3, Co_2O_3, MnO_2, Cr_2O_3, Ni_2O_3, and SiO_2, with a few mol% [10,11,12]. Such dopants play an important role on improving the microstructure of ZnO by modifying the grains size and grains growth during the sintering process which can in turn lead to improve the electrical properties. Recent research focuses on studying the factors that play an important role to make varistors more stable with low leakage current such as sintering temperature, the sintering period, and the type of sintering [13,14,15].

IOP Conf. Series: Materials Science and Engineering 1269 (2022) 012008 doi:10.1088/1757-899X/1269/1/012008

Annealing process is a heat treatment used to modify the microstructure to improve the crystallization and get rid of defects in material which involving slowly cooling the sample after heating at a specific temperature for a while to achieve a near equilibrium condition of characteristics [24]. Many researchers studied the effect of post annealing in different environments on improving the microstructure and homogeneity of grains and increasing their size, which inevitably affects the performance of varistors [1,2,14,15]. Besides, electrical and microstructure properties of varistors can be changed by modifying the initial particle size which enhance the grain size uniformity and the microstructure density [17, 18]. Bao-hui Chen and others showed that the properties of varistors improved in samples with particle size of 90 nm when they sintered at a temperature of 1100 °C for two hours [20]. Kunyong Kang and others investigated the effect of post-heat treatment on improving the structural of TiO_2-Nb_2O_5-$SrCO_3$ Ceramics varistors. Their results showed a significant improvement in the properties of the samples that were annealed in oxygen for 3 hours at 750°C with nonlinear coefficient (α) = 8.9 and break-down voltage (E_b) = 19.1 V·mm^{-1}[24]. Amal Boumezoued et al. found that the microstructure and electrical properties of ZnO–Bi_2O_3 varistors were improved when they annealed in a temperature range between 650 and 900 °C [25].

This work aims to study the influence of using different size of ZnO particles that mixed with little amount of metal oxides such as Bi_2O_3 and Mn_2O_3 to endorse the enhancement of the electrical behaviors and stability of ZnO based-varistors. This work includes studying the microstructure and electrical properties such as crystallinity, grain size, grain boundary, nonlinear coefficient, and break-down voltage of ZnO-Bi_2O_3-Mn_2O_3 varistors. In addition to, this work will show the effect of oxygen annealing process on ZnO based-varistors. Briefly, the experimental parameters investigated in this work showed significant impacts on both the structural and electrical properties of the devices.

2. Experimental details

For ZnO- Bi_2O_3- Mn_2O_3 varistors fabrication, four different particle sizes of zinc oxide powder: White ZnO with size of 0.23 μm, Pharma ZnO with size of 0.19 μm, ZnO nanoparticle with size 40nm, and 20nm, Bi_2O_3 and Mn_2O_3 with size of 1.0 ± 0.5 μm were purchased from Sigma-Aldrich Co. with purity of 99%. The conventional ceramic processing method was used to prepare ZnO-Bi_2O_3-Mn_2O_3 discs as follows: 99 mol% ZnO, 0.5 mol% Bi_2O_3 and 0.5 mol% of Mn_2O_3 mixed with polyvinyl alcohol (PVA) and distilled water in ball-milling jar for 6 hours. After mixing, they were dried for an hour in air at 60 °C. Then the ZnO- Bi_2O_3- Mn_2O_3 slurry was sieved through a 20-mesh sieve and granulated. Then, the mixed granulate was pressing at 4 ton/cm2 pressure to make discs of ZnO- Bi_2O_3-Mn_2O_3. The resulted green discs have a size of 28 mm in diameter and 2.5 mm thick. Finally, the green discs were then sintered for one hour at 1200 °C in air. To study the influence of oxygen, half of fabricated discs were annealed at 700 °C for 3 hours under oxygen atmosphere with flowing rate of 2 L/min.

2.1 Characterization

A scanning electron microscopy (SEM) (model: A JSM–6390LA) was used to study the microstructure of the ZnO- Bi_2O_3-Mn_2O_3discs. By using (a Jobin-Yvon T64000) triple spectrometer system, Raman scattering studies of ZnO powders were carried out in the backscattering geometry. All spectra were recorded by using the 514.5 nm line of a mixed argon/krypton ion laser at room temperature and in air.

2.2 Electrical testing

The varistors are prepared for electrical tests by applying a layer of silver paste on both sides of the discs and connecting them to the poles. Using a source measuring device (Keithley 2410, USA), the V-I characteristic of ZnO-based varistors was determined. The nonlinear coefficient (α) is identified by:

$$\alpha = \frac{logI_2 - logI_1}{logV_2 - logV_1} \quad (1)$$

where I_1= 0.1mA/cm^2 , I_2 = 1.0 mA/ cm^2 , and V_1 and V_2 are the electrical voltages corresponds to I_1 and I_2 respectively. The I-V curves' slope changes were used to calculate the breakdown voltages (from nonlinear to linear behavior).

3. Result and Discussion

3.1 Microstructural analysis

The microstructure of the White ZnO, Pharma ZnO, 40nm ZnO, and 20nm ZnO discs is illustrated in figure 1. In Figure. 1, the dominant constituent was ZnO grains which make up the bulk of the microstructure. The grains extremely similar concerning chemical composition. It is chemically composed of a large proportion of ZnO. In Figure 1, the structure is filled with the Bi-rich phases in the form of intergranular interlayers consisting mainly of bismuth, which largely accumulates between the grains and greatly affects the electrical conductivity of the varistors. The reason for this is that the melting temperature of bismuth is higher than the melting point of zinc oxide, and during the process of annealing, it condenses on the grain boundaries. Grain growth is affected by different additives, annealing time and annealing temperature. The increasing in sintering time and temperature increases the grain growth and increase its size [17, 18]. Furthermore, Figure 1 also showed that the grains are formed in polyhedral or hexagonal shapes, which is an ideal ZnO crystal [3]. All the samples have a homogeneous structure and low porosity, which plays an important role in the varistors' characteristics [4]. Table 1 shows the sizes of grains of the varistors discs. The grain size ranges from 9.47 μm to 15.11μm. Moreover, the samples showed a slight increase in the size of the granules, with a decrease in the size of the raw materials from 0.23 μm to 20 nm. The 20nm ZnO disc exhibited the largest grain size compared with the White ZnO samples. Because of the nanoparticles' high surface to volume ratio, this increasing in grains size may be explained [5,6]. The 20nm ZnO samples recorded the largest size of the grains by 15.11 μm, followed by 40nm ZnO disc by amount 13.21, as the increase in the size of the grains means an increase in the limits of grains, which inevitably affects the properties electrical varistors will be discussed in detail later, and this differs with the results obtained by Bao-Hui Chen and others, which stated that the size of the raw materials used in preparing the varistors did not affect the size of the grains [8]. Figure 2 illustrates the SEM micrographs of varistor samples, which were sintered at 1200 °C and annealed at 700 °C in the oxygen ambient. The figure shows an increase in structural homogeneity and regular grain growth and lack of porosity after annealing in oxygen, whereas annealing in oxygen increases the density of the receiving states and increases the potential voltage barrier at the grain boundaries, which improves the electrical properties of the varistors. Table 1 shows the comparing the average grains sizes before annealing in the ambient oxygen and after annealing, we find that a slight increase in grains sizes

(a)

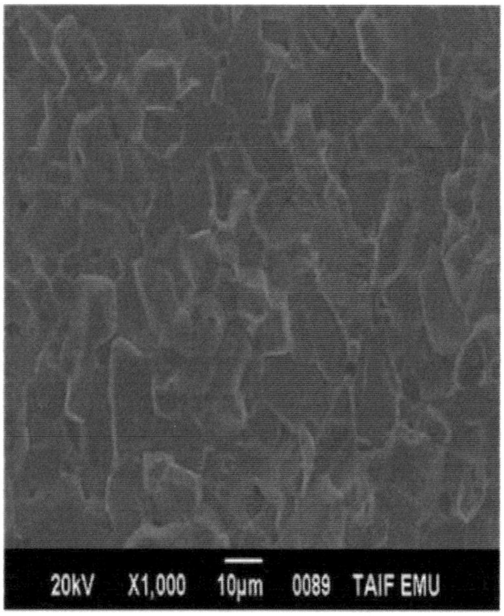

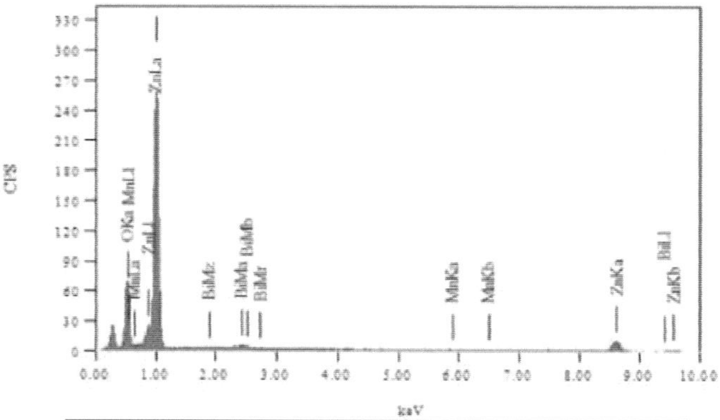

Element	Mass%	Atom%
O K	55.07	64.0462
Mn K	0.71	0.6084
Zn K	41.31	33.5517
Bi M	2.90	1.7937
Total	100	100

(b)

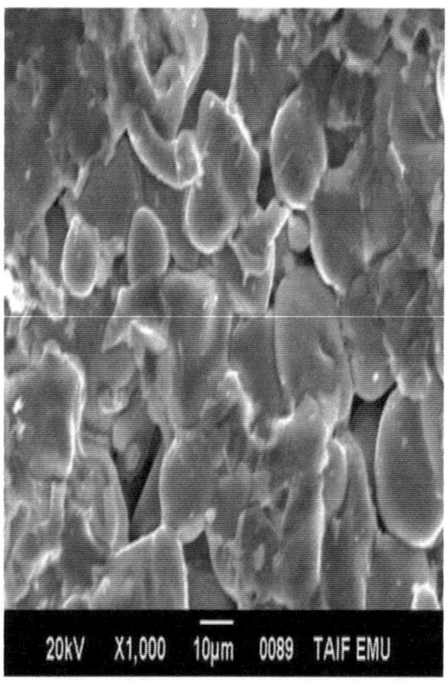

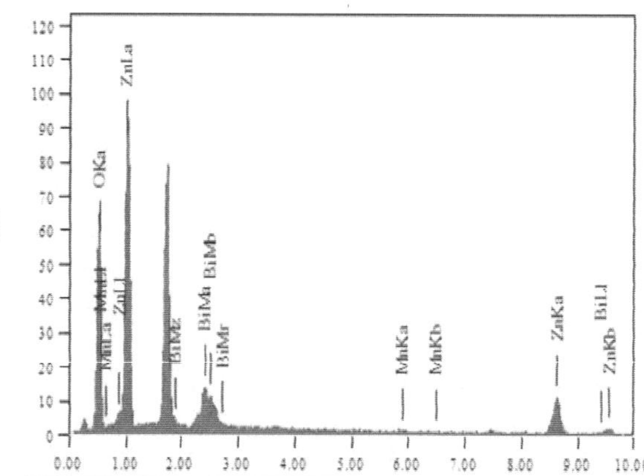

Element	Mass%	Atom%
O K	47.11	81.57
Mn K	0.32	0.16
Zn K	38.82	16.45
Bi M	13.75	1.82
Total	100	100

(C)

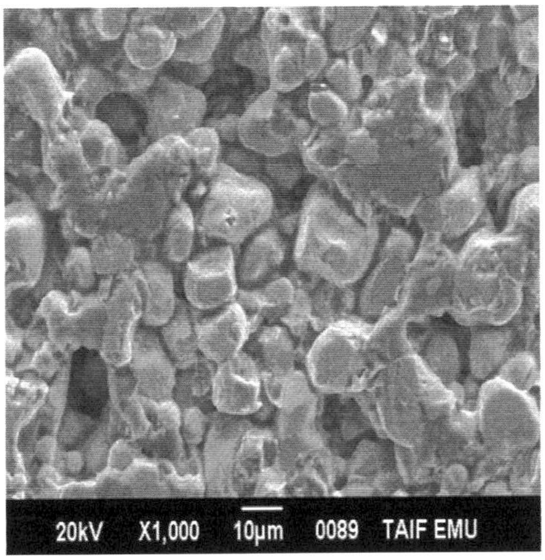

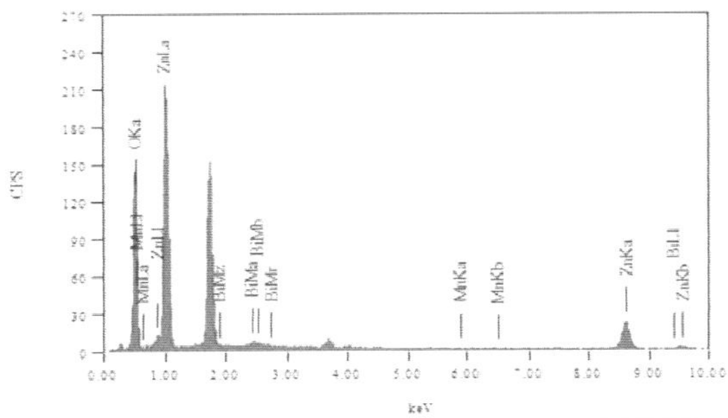

Element	Mass%	Atom%
O K	55.49	84.14
Mn K	0.46	0.20
Zn K	41.34	15.34
Bi M	2.71	0.31
Total	100	100

(D)

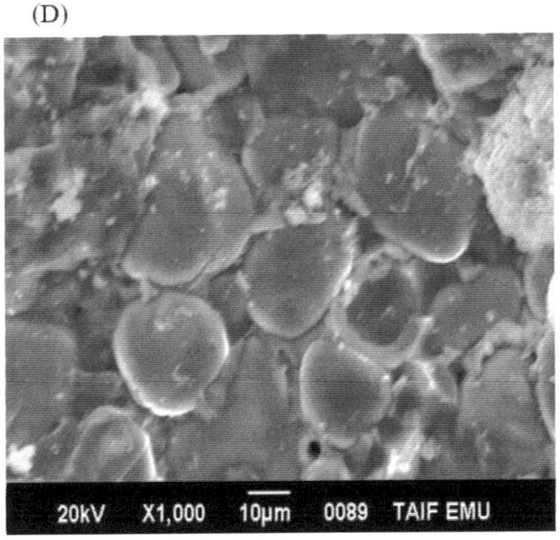

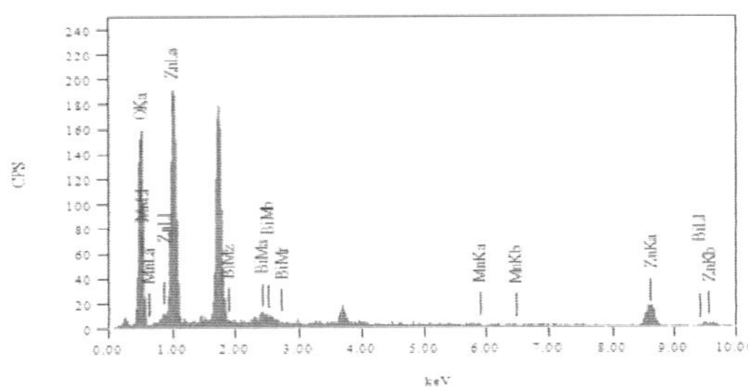

Element	Mass%	Atom%
O K	58.86	86.59
Mn K	0.33	0.14
Zn K	35.03	12.61
Bi M	5.78	0.65
Total	100	100

Figure 1. SEM-EDX analysis of ZnO- based varistor (a) White ZnO, (b) Pharma ZnO, (c) 40nm ZnO and (d) 20 nm ZnO

Table 1. Summary of a mean grain size of ZnO- based varistors without annealing and after annealing in O_2 atmospheres.

Samples	Mean grain size (μm)	Mean grain size after annealing in O_2 *atmosphere* (μm)
White ZnO	9.47	9.98
Pharma ZnO	12.70	12.99
40nm ZnO	13.21	13.55
20nm ZnO	15.11	15.44

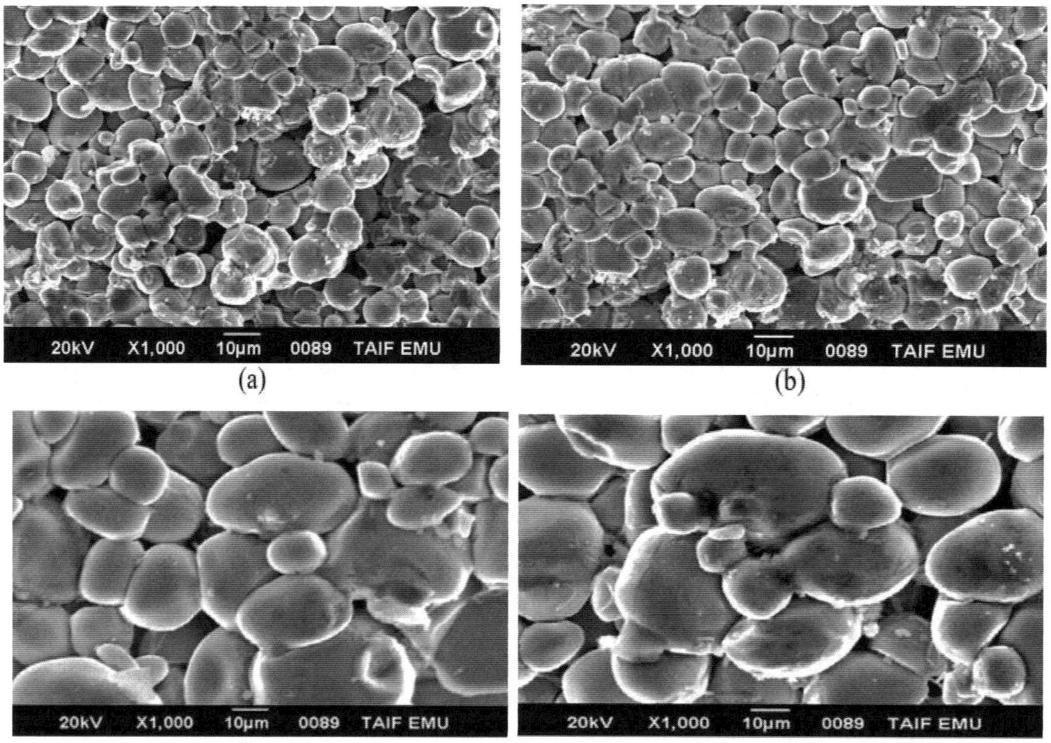

(a) (b) (c) (d)

Figure 2. Typical microstructure of ((a) White ZnO, (b) Pharma ZnO, (c) 40nm ZnO and (d) 20nm ZnO) discs after annealed at O_2 ambient.

3.2 Raman Investigation

Figure 3 displays the Raman spectra obtained from the White ZnO, Pharma ZnO, 40nm ZnO, and 20nm ZnO discs before and after annealing in Oxygen atmosphere. The space group C46v's hexagonal wurtzite structure, in which ZnO crystallizes, is formed. Eight sets of phonon modes are predicted by the group theory: $2E_2$, $2A_1$, $2B_1$ and $2E_1$. Among them, the $2B_1$ modes are not active by Raman spectrum. Table 2. illustrates the collect of the presented frequencies of Raman active phonon modes of the ZnO- based varistor comparison with the theoretical results. One of the most intensive modes in the Raman spectrum of all samples is a very tight (2 cm^{-1}) peak E_2low at 99 cm^{-1}, referred to as vibrations of the zinc sub-lattice in ZnO [22,23,24]. The other very intensive Raman mode is E_2^{high} at 437 cm^{-1} mostly referred to as the oxygen vibration [22,23,24]. Furthermore, the A_1 (LO) mode at 569 cm^{-1} and the E_1 (LO) mode at 586 cm^{-1} the presence of impurities and/or defects can influence both modes, the E_1 (LO) mode is more strongly affected by these effects [9]. The most intensive second-order mode in the low-wavenumber region, at 331 cm^{-1}, is ascribed to the difference E_2^{high}- E_2^{low} [9]. The very large shift in the position of the E_2 (high) mode of all samples comparison with the theoretical results of indicates that the addition of Bi_2O_3 and Mn_2O_3 strengthens the tensile stress of the ZnO crystals. In general, substituting Zn^{2+} by Bi^{3+} in a ZnO crystal is complicated due to the radius of Bi^{3+} is substantially larger than that of Zn^{2+}[10,27]. In general, this red-shift in E_2(High) is due to defects and internal strains due to Bi_2O_3 segregating fundamentally at the grain boundaries as bi-rich phases during the sintering process, and this promotes grain growth and increases the stress [11]. Also, the lattice distortion is resulting substitution of the dopant ions, such as Mn^{2+} and Mn^{4+} for Zn^{2} during the sintering [10]. The bands of the secondary phases, such as b-Bi_2O_3 and Mn_3O_4, were noted in the Raman spectra. The position of the b-Bi_2O_3 Raman peak is at approximately 332 cm^{-1} and that of Mn_3O_4 is at 574 cm^{-1}. The peak at 540 cm^{-1} is more dominant in the case of annealing in oxygen for all samples.

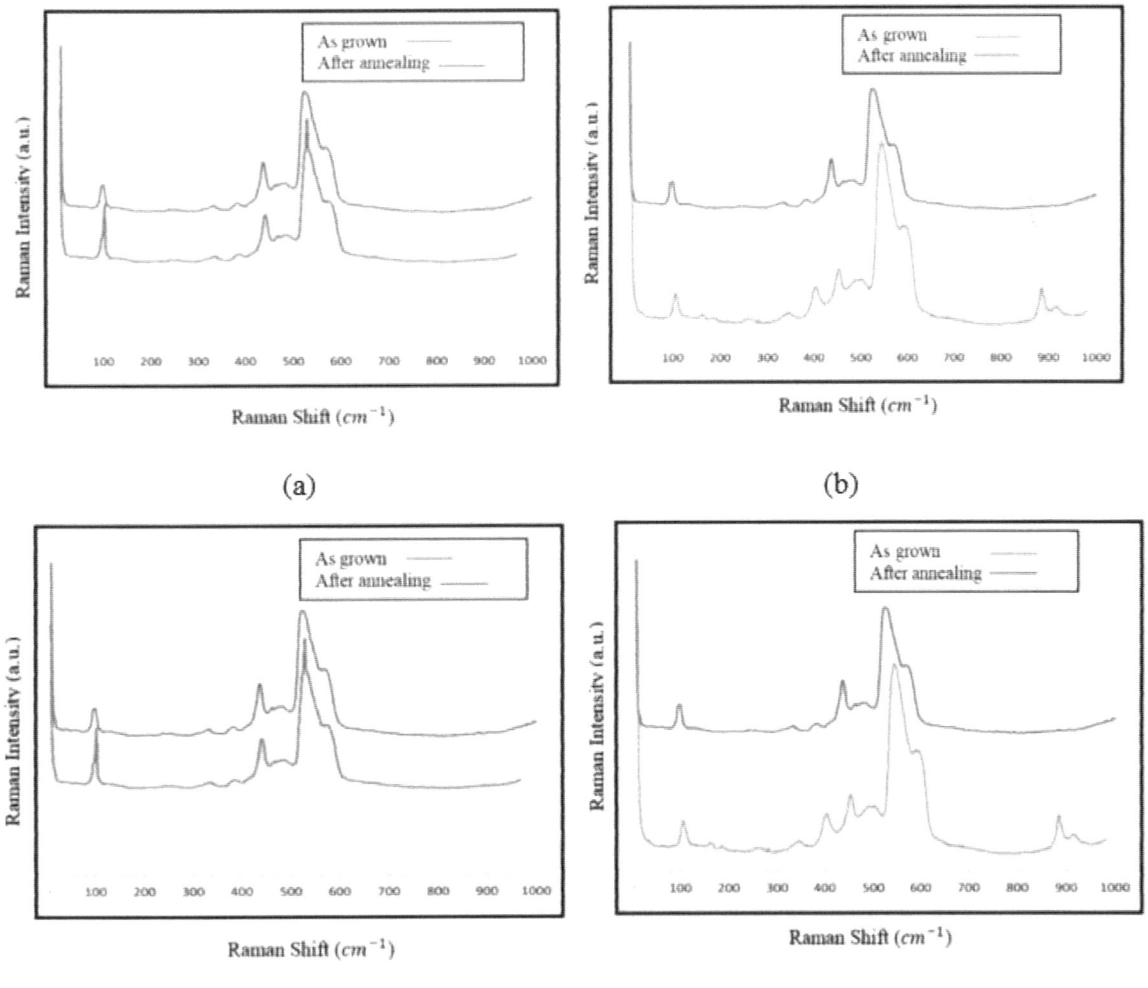

Figure 3. Raman spectra of ZnO- based varistor made from different ZnO particle size and after annealing in O_2 ambient (White ZnO, (b) Pharma ZnO, (c) 40nm ZnO and (d) 20nm ZnO) discs.

Table 2. A comparative of the active modes of Raman spectra of the ZnO- based varistors manufactures from micro and nano ZnO powder with the theoretical results before and after annealing in O_2 ambient.

Ref [24]	Symmetry	Process	White ZnO		Pharma ZnO		40nm ZnO		20nm ZnO	
			ZnO-based varistor	After annealing	ZnO-based varistor	After annealing	ZnO-based varistor	After annealing	ZnO-based varistor	After annealing
101	E_2	E_2^{low}	99	101	100	99	100	99	99	99
208	$A_1, (E_2)$	2TA;2E_2^{low}	-	-	-	-			272	270
332	$A_1, (E_2, E_1)$	E_2^{high}-E_2^{low}	334	334	325	320	333	333	333	333
380	A_1	$A_1(TO)$	380	380	386	390	385	387	385	389
408	E_1	$E_1(TO)$	-	-	-	-	-	-	-	-
437	E_2	E_2^{high}	431	433	430	432	432	431	430	431
541	A_1	2B_1^{low};2LA	520	528	527	529	524	527	523	520
574	A_1	$A_1(LO)$	576	575	575	579	574	570	574	-
584	E_1	$E_1(LO)$	-	-	-	-	-	-	-	590

3.3 Electrical properties

Figure 4 shows the I-V curves ZnO-based varistors which made from different size ZnO powders before and after annealing in O_2 ambient. Table 3 determines the nonlinear coefficient and breakdown voltage of the samples. Where, it becomes clear that the breakdown voltage increases with the decrease in the particles size of materials used in the manufacture of the varistors, where the sample 20nm recorded the highest breakdown voltage of 2555 volt. In contrast, samples White ZnO recorded the lowest breakdown voltage of 201 volts. It has been proved by Chul- Hong Kim and C–C Lin. the breakdown voltage is dependent on the particles size of raw materials [26]. With annealing in O_2 ambient, the breakdown curve shifts to higher electric voltage regions. Where the increase in breakdown voltage for sample White ZnO was 11.9% and the sample Pharma ZnO achieved an increase of 14.45 % and 14.59 % and 8.10% for 40nm ZnO and 20nm ZnO disc respectively. Figure 4 shows The higher breakdown voltage with decreasing size of the raw materials in the samples 40nm ZnO and 20nm ZnO discs is due to the properties of the materials in the nano size due to the large surface area relative to the size of the nano-powders, as in the case of higher grain size, the area of the grain boundaries increases, which acts as depletion areas of the current as the Fermi level of the grain is higher from the grain boundary level causing free electrons to flow into the grain boundary to achieve equilibrium and are trapped by the defects and doping in the grain boundary material to become a negative interface charge and become the grains act as insulators in the way of current [14]. As previously proven in SEM results, the decrease in the volume of raw materials used in the manufacture of varistors led to an increase in the size of the grains and consequently an increase in homogeneity and lack of porosity, which is related to an increase in the non-linear coefficient of varistors, which means improved electrical properties. Where the sample of 20nm ZnO achieved the highest nonlinearity coefficient at 49.89 as illustrates in table 3. The decrease in porosity related to the increasing in the nonlinear coefficient as has been proved in another study [15].

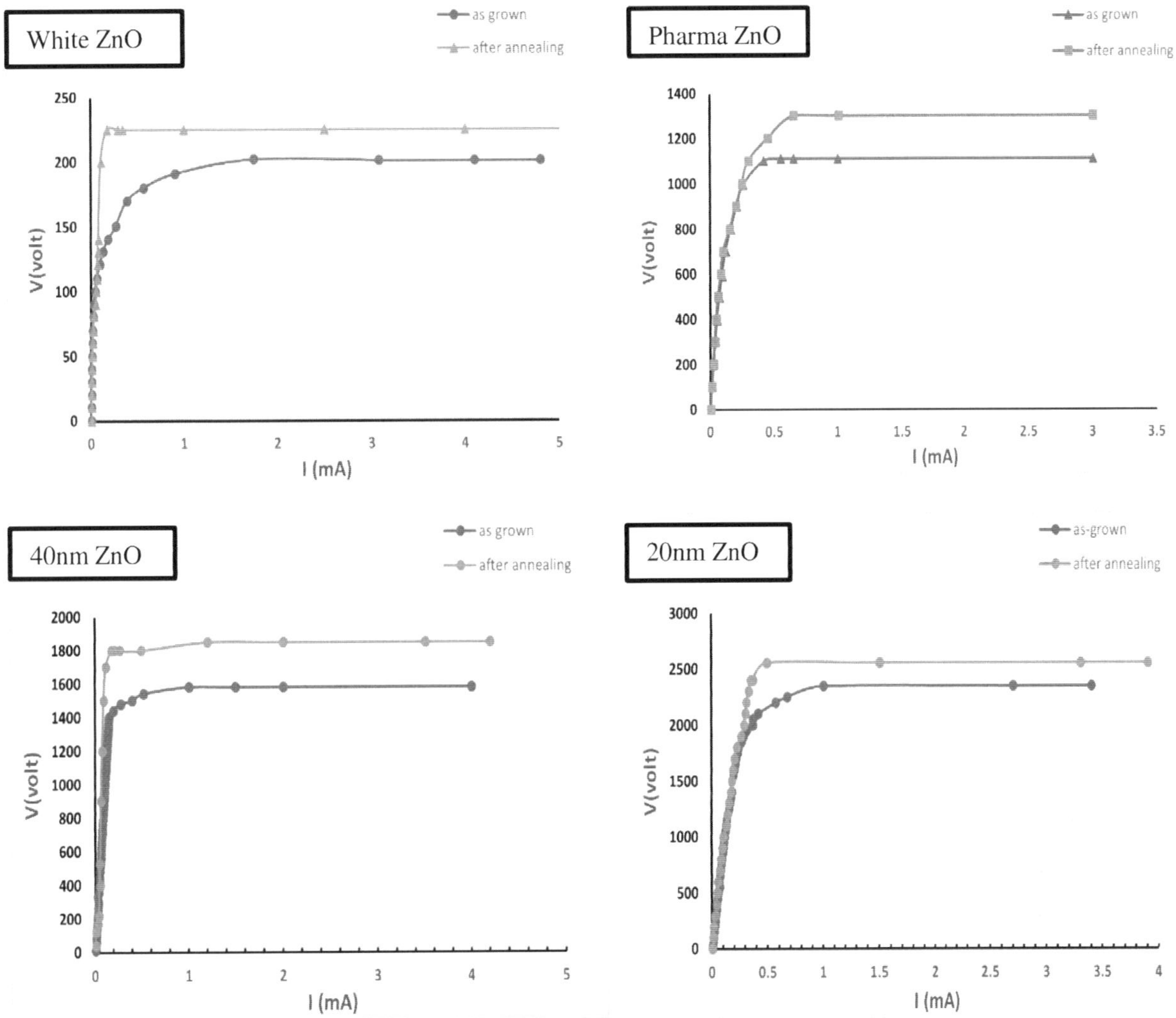

Figure 4. I-V characteristics of ZnO- based varistor made from different ZnO particle size as grown and after annealing in O_2 ambient (White ZnO, (b) Pharma ZnO, (c) 40nm ZnO and (d) 20nm ZnO) discs.

IOP Conf. Series: Materials Science and Engineering 1269 (2022) 012008 doi:10.1088/1757-899X/1269/1/012008

Table 3. Summary of the electrical properties of ZnO- based varistors made from different ZnO particle size before and after annealing in O_2ambient.

Sample		V_b(V)	α
White ZnO	ZnO- based varistor	201	32.5
	After annealing	225	33.4
Pharma ZnO	ZnO- based varistor	1111	33.3
	After annealing	1300	37.19
40nm ZnO	ZnO- based varistor	1580	43.6
	After annealing	1850	47.12
20nm ZnO	ZnO- based varistor	2348	44.5
	After annealing	2555	49.89

4. Conclusion

In this study, the influence of different ZnO particle sizes and the oxygen annealing process on the microstructure, and electrical properties of ZnO-based varistors was investigated. Have obtained these interesting results, when the particles size of zinc oxide was 20nm; the best structural homogeneity and electrical properties was obtained, as this sample recorded a breakdown voltage of 2555 volt and an improved nonlinear coefficient of 49.89 because the improved grains uniformity. Also, the annealing in oxygen can improve the microstructure and electrical properties of ZnO-based varistors. The enrichment of oxygen at grain boundaries increases the height of potential barrier at boundaries of grains and the density of acceptor state, therefore α increase. And it was concluded that the varistors made of zinc oxide nano-powders have better properties than those made of powder with a larger size material. And it was concluded that the varistors that were annealed in the oxygen environment have better properties than those that were un-annealing in oxygen.

References

[1] Wang et al X 2021 Eur. Ceram 41(4) 2633-2640.
[2] Rohini R, C Pugazhendhi Sugumaran 2021 Electr. Eng. Technol 16(1) 469-481.
[3] Zhang L, Gao L, Liu W, Guo Q, Li S, Li J 2021 J. Eur. Ceram 41(4) 2641-2647.
[4] Hembram K, Rao T, Ramakrishana M, Srinivasa S R, Kulkarni R A 2020 J. Alloys Compd 817 152700.
[5] Huang Y, Guo M, Li J 2020 Ceram. Int 46(14) 22134-22139.
[6] Cui F, Xu Z, Chu R, Li G 2020 J. Alloys Compd 836 154692.
[7] Li J, Yang S, Pu Y, Zhu D 2021 Mater Sci Semicond Process 123 105529.
[8] Shi et al M 2022 J. Mater. Sci. Mater. Electron 33(24) 19242–19251.
[9] Hamdelou S, Guergouri K, Arab L 2015 Appl. Nanosci 5 (7) 817–825.
[10] Y Yan et al 2020 J. Am. Ceram. Soc 103(5) 3265–3272.
[11] Hembram K, Rao N T, Srinivasa S R, Kulkarni R A 2015 J. Eur. Ceram 35(13) 3535–3544.
[12] Zhao Y Z, Wang -H M, H.-P. Zhang -P H 2016 J. Mater. Sci. Mater. Electron 27(2) 1777–1782.
[13] Roy S, Roy K T, Das D 2019 Ceram. Int 45(18) 24835–24850.
[14] Ortiz-Mosquera F J, Nieto-Munoz M A, Rodrigues M C A 2021 J. Alloys Compd 850 156670.
[15] Zhang Y, Colella R, Kycia S, Goldman A L 2002 Acta Crystallogr. Sect. A Found. Crystallogr 58(4) 385–390.
[16] Wong J 1980 J. Appl. Phys. 51(8) 4453-4459.
[17] Olsson E, G. Dunlop, Österlund R 1993 J. Am. Ceram. Soc 76(1) 65-71.
[18] Gupta K T, Carlson G W1985 J. Mater. Sci 20(10) 3487–3500.

[19] Sendi K R 2015 university sains m,Alaysia 4452-3323.
[20] Chen B, Wang B, Gao P, Zhang P, Chen H 2021 J. Alloys Compd 872 159638.
[21] Alim A K, Fonoberov A V, Shamsa M, Balandin A A 2005 J. Appl. Phys 97(12) 124313.
[22] Luo L, -M Y. Chiang 1999 J. Eur. Ceram. Soc 19(6–7) 697–701.
[23] J. M. Calleja and M. Cardona1977 Phys 1977 Rev. B 16(8) 3753.
[24] Kang, Kunyong, et al 2021 Materials Sciences and Applications 12 461-474.
[25] Boumezoued, Amal, et al 2021 Journal of Materials Science: Materials in Electronics 32(3) 3125-3139.
[26] Kim -H C, Kim -J H 2004 J. Eur. Ceram Soc 24(8) 2537–2546.
[27] Sendi K R, Munshi A, Mahmud S, 2014 Superlattices Microstructure 69 212–225.

Titanium Dioxide (TiO_2) Concentration-dependent Photovoltaic Cells Performance of PAni-TiO_2 Nanocomposite

M. Abdelhamid Shahat *, Ahmed Ghitas

PV Unit, Solar and Space Research Department, National Research Institute of Astronomy and Geophysics (NRIAG), 11421 Helwan, Cairo, Egypt.

* **Corresponding Author** E-mail: m.abdelhamid999@gmail.com

Abstract. By optimizing the features of the photoactivated polyaniline-titanium dioxide (PAni-TiO_2) nanocomposite layer, varied concentrations of TiO_2 nanoparticles (NPs) inside PAni matrix were used to improve the performance of organic cells. In this work, polymer solar cells were designed using the FTO/(PAni-TiO_2)/Ag combination. PAni-TiO_2 films' structural evolution, surface properties, optical and electrical characteristics have been studied. The XRD patterns showed that the crystallite size gradually decreases from 18.7 to 12.8 nm depending on the TiO_2 NPs concentration in the PAni matrix. In addition, the surface of the highly concentrated films was rougher and more porous than it was for the pristine film. With varying concentrations of TiO_2 NPs incorporated into the PAni structure, PAni-TiO_2 based composite cells' efficiency enhanced from 0.33 to 0.85%. This conclusion is mostly related to a structural shift accompanied by a high increase in roughness scales, that led to a reduction of reflected photons and, as a consequence, an increase in the creation of free carriers.

Keywords: Polyaniline (PAni), PAni-TiO_2 Nanocomposite, Energy Band Gap, Electrical conductivity, Organic Solar Cells,

1 Introduction

Organic solar cells have gotten a lot of interest from scientists because of its high photoelectric conversion efficiency, easy fabrication technique, and inexpensive production costs. In general, metal oxides with large surface area play the main role in improving the performance of nano nanocomposite photovoltaic cells. Sequentially, titanium dioxide (TiO_2) NPs, in particular, offer appealing characteristics such as large surface area, easy fabrication, low cost and great chemical stability [1]. Further, it's are ideal for a wide range of applications, including PV cells, water desalination, Photocatalytic uses and electronics [2,3]. Besides, it can also be used in energy storage, biosensors and piezoelectricity [3,4].On the other side, owing to their excellent physical and electrical characteristics, conjugated polymers are extensively employed in a variety of tenders [5]. Polyaniline's (PAni) conductivity changes from semiconductor to metal conductivity, contrasting other conductive polymers like polypyrrole and polythiophene, and it can be readily regulated by adjusting the incorporating ratio [6]. PAni also has a number of other appealing characteristics, including a optimistic temperature factor of resistance, photocatalytic qualities, ease of and low cost of synthesis, and sensitivity to chemical. As a result, PAni has applications in solar cells and anticorrosion coatings [6, 7]. Inorganic materials composition and conducting polymers have lately concerned a lot of consideration as advantages that may be used to produce high-quality final nanocomposite [8]. Previously investigations have shown that the chemical characteristics of PAni-TiO_2 composites change significantly from those of unmodified PAni and TiO_2, owing to the two materials chemical interaction [8]–[10].

PAni-TiO_2 nanocomposites have a high electrical conductivity compared to their respective components [11]. Furthermore, in acidic media solvents, TiO_2 molecules can collect substantially, reducing both the degree of absorption and the surface area of the TiO_2-polymer interface. Markus Diantoro et al. were designed an organic dye-sensitized solar cells (DSSC) based-TiO_2-PAni/ITO and its performance was enhanced using different PAni concentrations [12]. Kailash Nemade and et al., were investigated different amount of TiO_2 NPs in the composite system between PAni-graphene/TiO_2 for DSSC applications. Besides, the performance of fabricated device was enhanced to the optimum conversion efficiency at 1 wt% of TiO_2 NPs [13].

Herein, chemical polymerization was used to produce PAni-TiO_2 nanocomposites by incorporating the TiO_2 NPs in the PAni matrix at various concentrations. The PAni-TiO_2 composites are employed as an active photovoltaic layer in a cell structure. The components of a cells have been created using a variety of chemical and physical approaches. Additionally, J-V curves of the PAni-TiO_2 photo-layers are performed in simulated sunlight with a full illumination intensity of 100 mW/cm^2 to investigate the ratio outcome of NPs within the PAni structure on the efficiency of solar cells.

2 Experimental procedures

2.1. Reagents and Materials

For the synthesis of TiO_2 NPs, titanium dioxide powder (Ranbaxy, > 5m) was utilized as row materials. Aniline monomer (C_6H_7N, Purity 99+ %, Germany), hydrochloric acid HCL (37%), ammonium peroxydisulfate (APS) ($(NH_4)_2S_2O_8$), Germany), and distilled water were used to produce PAni. Sigma Aldrich provided all of the chemical materials.

2.2. Synthesize of PAni-TiO_2 composites and fabrication cell devices

TiO_2 NPs were synthesized chemically using a hydrothermal method [1]. To prepare PAni-TiO_2 composites, different weight of TiO_2 nanopowders (20, 25, and 33%) were mixed to the unmodified PAni solution during the chemical oxidative polymerization process of aniline monomer [14]. Then, the synthesized composites of PAni-TiO_2 were deposited via a spin-coater device on FTO (81-84% transmittance and 4.1 x 10^{-5} Ω.cm resistivity, sigma Aldrich) coated conducting substrates using an 800 rpm for 30s. Moreover, A complete solar cell system based on the as prepared photoactive film is obtained just by depositing-a highly conductive cathode of silver (Ag) paste (high concentration ≥76%, low resistivity of 1-3.1 x 10^{-5} Ω.cm) on the photoactive surface using a screen printer machine (EQ-SPC-2-LD). Fig. 1 displays a design of the stack device architecture.

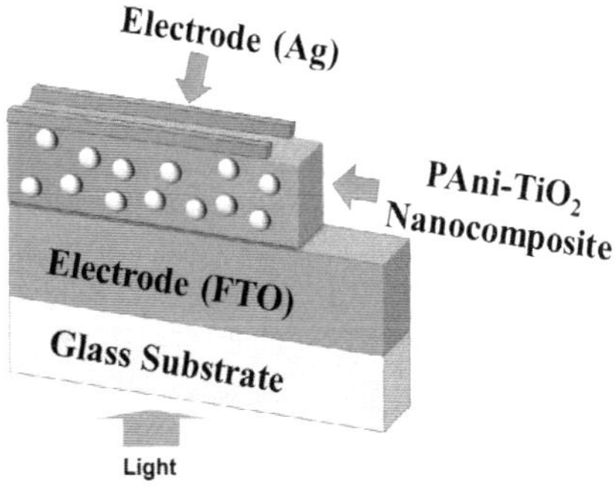

Figure 1: Device architecture of PAni-TiO_2 nanocomposites heterojunction device.

3 Characterization Techniques

Using XRD, the generated samples' phase purity and structure have been examined (Bruker Model D8 ADVANCE). Further, using FTIR Spectroscopy (Jasco Model 4100 Japan) the functional groups of nanocomposite films are analyzed. The morphology of the generated unmodified and modified PAni layers was examined using a high-resolution SEM (JOUL, JSM-5500, Japan) at an accelerated voltage of 10 KV. Additionally, SPECORD 200 PLUS spectrophotometer was used to assess the optical characteristics. Besides, a profilometer was used to measure the surface roughness (Ra) (Talysurf 50-Taylor Hopson precision). Meanwhile, an EQ-JX2008-LD resistivity tester was used to evaluate electrical resistivity using the four-point probe (4PP) technique. Each film's average electrical resistivity was estimated using 10 readings collected from various places. Moreover, a solar simulator system is used to test the J-V properties of the solar cell under a standard light condition of 1.0 Sun at 25 °C. The simulated test was performed utilizing a xenon lamp as a light source with fixed conditions (AM 1.5 G, 100 mW/cm2) using a KEITHLEY 2400 Source Meter.

4 Phase structure and composition study

4.1 X-Ray diffraction analysis

At various TiO_2 NP concentrations incorporated into the PAni matrix, the structural characteristics of PAni-TiO_2 nanocomposites were determined using the XRD spectrum (see Fig. 2). In contrast to other amorphous PAni materials, the XRD pattern proves sharp, well-defined peaks, confirming the high crystallinity of the pristine PAni [15]. Herein, The TiO_2 peaks in this study agree well with the tetragonal structure with anatase phase ($a = b = 3.7335$ Å, $c = 9.3788$ Å, $\alpha = \beta = \gamma = 90.0$) (ICSD, Code no 9852) [16]. Meanwhile, the orthorhombic structure is corresponded to the crystalline unmodified PAni ($a = 7.3574$ Å, $b = 6.9675$ Å, $c = 10.1714$ Å, $\alpha = \beta = \gamma = 90.0$) [17], [18]. Different TiO_2 concentrations are used to create polycrystalline diffraction patterns in PAni-TiO_2 compounds that are comparable to those produced by typical powder diffraction patterns for the tetragonal anatase TiO_2 structure and the orthorhombic structure of PAni. This suggests that there isn't any phase shifting going on throughout the incorporating process.

Due to the polymer chains being positioned parallel and adjacent to one another, crystalline zones are produced, which are surrounded by amorphous regions, and the diffraction peak occurs at 2θ from 23° [19]. Additionally, the incorporating of TiO_2 NPs has a significant impact on the peak intensity of the composites and this affect weakens with increasing NPs ratio. These outcomes might be explained by an improvement in the physicochemical interaction between PAni and TiO_2 NPs as the quantity of NPs increases. Besides, the monomer fragments that are included in the low density can be linked to the formation of TiO_2 grains being inhibited, which in turn slows down the movement of contaminated particles on the growth surface and causes TiO_2 to crystallize [20]. Similar effect was seen in chemically fabricated PAni-TiO_2 composites, where the PAni matrix inhibited the formation of nanocrystalline anatase TiO_2 NPs [21, 22].

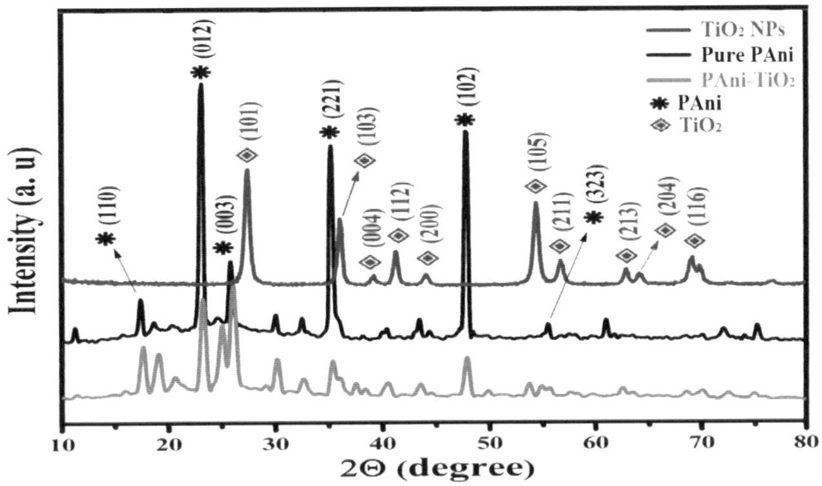

Figure 2: XRD spectra of TiO_2 NPs, pristine PAni and nanocomposites of PAni-TiO_2.

The resultant composites' crystallite size was subsequently determined using Williamson-plot Hall's equation (1). (See Table 1).

$$\Gamma\ cos\theta = \frac{0.94\lambda}{D} + 4\varepsilon\ sin\theta \qquad (1)$$

where $\Gamma(2\theta)$ is the integral peak breadth formed by the microstrain of the investigated material at diffraction angle of 2θ in radians, λ is the X–ray wavelength, D is the average crystallite size, and ε is the residual microstrain [23]. As noted, as the ratio of TiO_2 in the matrix structure increases, the size of the crystallite gradually decreases.

Table 1. Crystallite size and Microstrain of pristine PAni and composites of PAni-TiO_2 synthesized at different concentrations of TiO_2 NPs

Sample	**TiO_2 NPs Ratio**	**Williamson-Hall**	
		D size (nm)	Microstrain X 10^{-4}
PAni-TiO_2	0%	18.73	9.3
	20%	15.29	20.5
	25%	14.43	20.3
	33%	12.89	7.3

4.2 FT-IR spectra

So as to describe the structure and interaction formulation bonding characteristics of the films as-prepared, FT-IR analysis was performed. Fig. (3) compares the FT-IR spectra of modified PAni with TiO_2 nanopowders in various ratios (0, 20, 25, and 23%) obtained during a reaction time of 8 h to those of unmodified PAni. The peaks are essentially present at 3121 cm^{-1},1627 cm^{-1}, 1537 cm^{-1}, 1412 cm^{-1}, 1193 cm^{-1}, 860 cm^{-1}, 616 cm^{-1} and 473 cm^{-1}, as can be seen in the spectra of all the samples. The stretching of the hydrogen-bond O–

H and the stretching of the aromatic amine's N–H bond are responsible for the key distinctive bands that were discovered between 3690 and 2983 cm^{-1} [24], [25].

Hydroxyl radicals can develop on the surfaces of hydroxyl bonds when photo - induced holes are accepted [26]. These radicals can thereby enhance the composite's photocatalytic activity [27]. Further, in contrast to that observed in pure PAni, the peak is located at 3121 cm^{-1} in a composite that is distinguished by strong intensity, varied stretching form, and a little movement near the low wavenumber. This could be explained by the attraction of two close functional groups or a cluster of nearby atoms [28]. Additionally, the peak at 1627 cm^{-1} is related to the stretched Ti-O bond [29]. Meanwhile, the C=N and C=C stretching vibrations of the quinonoid and benzene rings are thought to be responsible for the 1537 cm^{-1} and 1412 cm^{-1} peaks, respectively. The C=N and C=C peaks in the spectrum of PAni-TiO_2 composites, which are the main peaks of PAni, provided evidence that PAni had formed on the nanoscale TiO_2 surface [30]. Moreover, the peak of the PAni conductivity at 1193 cm^{-1} is caused by the PAni quinonoid unit, which is employed to evaluate the delocalization degree [31]. The vibrations in the stretched Ti-O-Ti bonds are responsible for the other, which has a broad peak between 800-450 cm^{-1} [32], [33]. The benzenoid unit's C-C and C-H are linked to the detected peaks at 860 cm^{-1} and 616 cm^{-1} [34]. Whereas, the peak at 616 cm^{-1} is dramatically higher than it is for unmodified PAni, this could be related to vibrations in TiO_2's stretching Ti-O-Ti bonds and out-of-plane bending flexural vibration of PAni's C-H aromatic rings [24].

The incorporation of TiO_2 NPs into the PAni structure as a result of a powerful H-bonding between PAni and TiO_2 NPs results in a small peak shift towards higher wavenumbers, however the pristine PAni main peaks remain to appear obviously in a composite structure in the area 3121–473 cm^{-1}. Besides, the presence of TiO_2 NPs is responsible for the increase in peak area relative to PAni, which denotes a changes in the PAni structure. The appearance of TiO_2 peaks next to pristine PAni peaks indicates that PAni-TiO_2 composites have formed. According to coupling between the N atoms in the PAni chains and the TiO_2 NPs, the PAni layer able to effectively interact with the TiO_2 surface and cover the nanoparticles [35], [36].

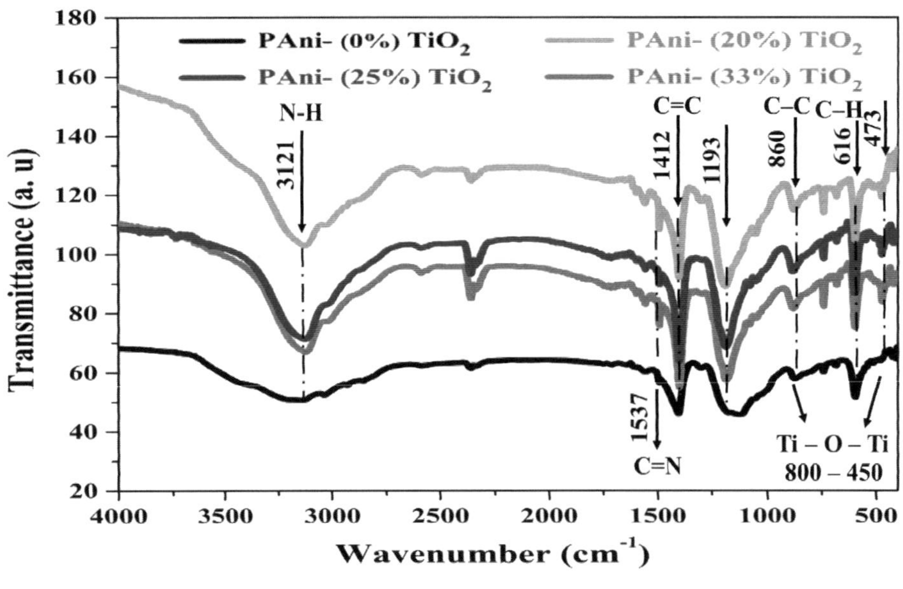

Figure 3. FT-IR spectra of unmodified PAni and PAni-TiO_2 layers prepared with different ratios of TiO_2 NPs.

4.3. Morphological Study

The TiO_2 NPs are characterized with TEM image (See Fig. 4a) a semi-spherical particle form with some particle size variations. After adding TiO_2 NPs in a variety of concentrations to a PAni matrix, the composites exhibit

deformed chain structure, structural characteristics, and porous topography that can be clearly seen in Fig. 4. This may be attributable to the long PAni chains being evenly distributed and the development of a PAni shell on the surface of the TiO_2 NPs, which creates repulsive forces between the nanoparticles and prevents their agglomeration [37], [38]. As a result, nanoflakes are used to form a morphological structure [39]. According to earlier research's, developing active layers with a microporous form result in a surface that is rather rough, this improves the composite layers' surface contact with the electrode surface [40]. The composite's active layer's microporous surface is where the composite's higher capacitance can occur due to maximal absorption [39]. The morphology of the composite films is influenced by the polymerization conditions, including pH, temperature, polymerization duration, and the structure and ratio of TiO_2 [41]. Additionally, the white region is designated as TiO_2, whilst the grey and black parts are associated with the PAni chain. The PAni-TiO_2 layers shown enhancement in the surface distribution of white area in comparison to the unmodified sample by increasing the NPs concentrations from 0 to 33% (Fig. 6b-g). This might prove that, with higher TiO_2 incorporation, the mixed TiO_2 is dispersed uniformly throughout the surface of the PAni. Similarly, sample with a high concentration ratio (33%) showed the creation of nanoflakes in the composite structures [40], [39].
Additionally, the enhancement in the physicochemical interaction between the TiO_2 nanopowders and PAni is responsible for these morphological phenomena. Finally, it can be said that an increase in the ratio of TiO_2 NPs in composites resulted in a rise in:

- The PAni matrix's adsorptive qualities
- The capacity to regulate the nanocomposite's structure
- The area of nanoflakes' surfaces

The effectiveness of the electrical properties and photocatalytic activity are improved by these nanoflakes [42].

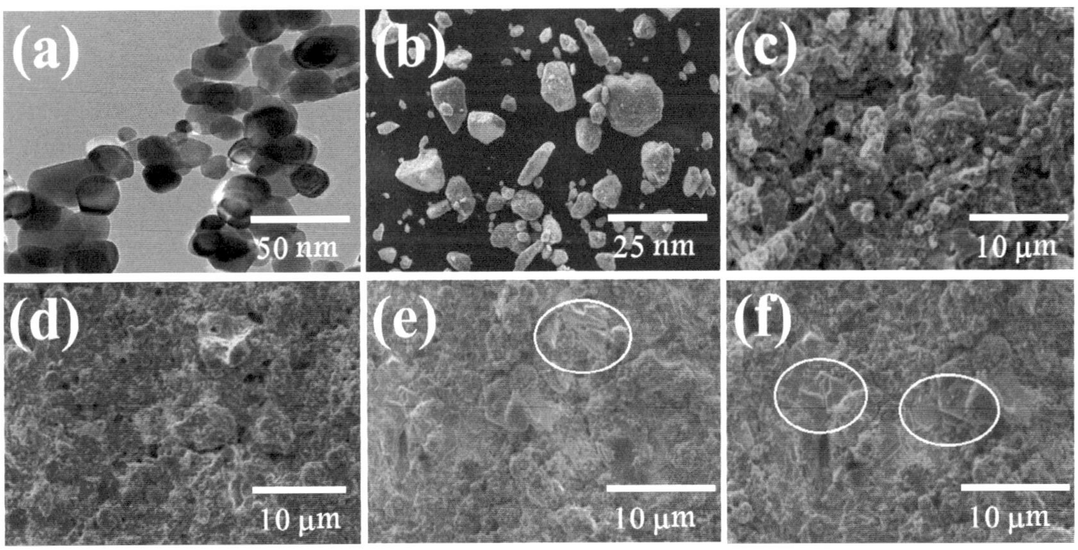

Figure 4 (a-f) TEM image for pure TiO_2 NPs, SEM images for pure TiO2 NPs and PAni-TiO_2 nanocomposite at ratio of (0, 20, 25, 33) %, respectively; the TiO_2 NPs prepared at reaction time of 8h.

4.4 Roughness Measurements

Surface roughness measurements are affected by the incorporation of TiO_2 NPs at different concentrations into PAni matrix, which also reveals a wide variety of nanofilms (see Table 2). With the addition of 20% and 25% of nanopowders, the average roughness value (Ra) was discovered to have decreased from 2.5 m of unmodified PAni to 1.91 µm and 1.48 µm. An optimum roughness value of 2.61 m was then attained with a higher TiO_2 NP concentration (33%). The structural features differences between the composites are mostly responsible for this trend's interpretation. Surface roughness is enhanced as a result of interactions of the molecules, the TiO_2 NPs' size inside the PAni structure, and their distribution across the lengthy PAni chains [43]. however, surface smoothness is mostly provided about by the fibrous network structure. It is well known that the optical characteristics of nanocomposite films, such as transmittance, absorbance, and optical constant, are affected by changes in surface roughness [44].

Table 2. Roughness factor (R_a) of PAni-TiO_2 nanocomposites prepared at 8h with different ratios into polymer matrix.

TiO_2 ratio in polymer matrix	Ra (μm)
0 %	2.51
20%	1.91
25%	1.48
33%	2.61

4.5 Optical properties

Investigation and discussion were performed on the optical characteristics of unmodified and modified PAni nanofilms as a function of TiO_2 ratio. The absorbance and transmittance spectra of each produced sample were measured at close to normal incidence across the wavelength range of 335-800 nm, as shown in Fig. 5. For both pure and modified PAni with absorption peak shifts, Fig. 5a displays a comparable spectrum with wide absorption peaks in the wavelength ranges of 335-465 nm and 550-800 nm. As the concentration of TiO_2 NPs increases, the 550–800 nm absorption band moves towards shorter wavelengths. This band shift shows the impact of physiochemical reactions on the electronic structure of composite films, which may interact with the valence band of TiO_2 NPs and the higher occupied molecular orbital (HOMO) of PAni [42]. At wavelengths of 330, 420, and 600–800 nm, unmodified PAni exhibits an energy absorption equivalent that, in turn, causes transitions of $\pi \rightarrow \pi^*$ inside the benzenoid rings, Polaron→π^* of the quinoid rings, and π→polaron excitonic of the quinoid rings, respectively [45]. These detected peaks in the PAni-TiO_2 nanocomposites eventually intensify with a small shift toward higher wavelengths. This demonstrates that the addition of TiO_2 NPs had a significant impact on the polaron tape's production in the PAni bandgap. For pure PAni, a peak that corresponds to the excitonic transition between benzenoid and quinoid is also seen at 690 nm. For PAni-TiO_2 (0%) nanofilm, a minimum absorption peak is mostly seen. Subsequently, as the nanopowders concentration ratios grew, absorption gradually increased until it reached the optimal value at 33%. This is due to the PAni and TiO_2 NPs favorable interactions [46]. Previous research has shown that the PAni-TiO_2 absorption curve has the similar characteristic [47], [48]. On the other hand, the TiO_2 NPs ratio has a significant impact on the transmittance spectrum (see Fig. 5b). When 33% of TiO_2 NPs were introduced into the PAni composition, the least transmission was seen. This correlates to the UV's π-π^* transition and the visible areas' bipolaronic band.

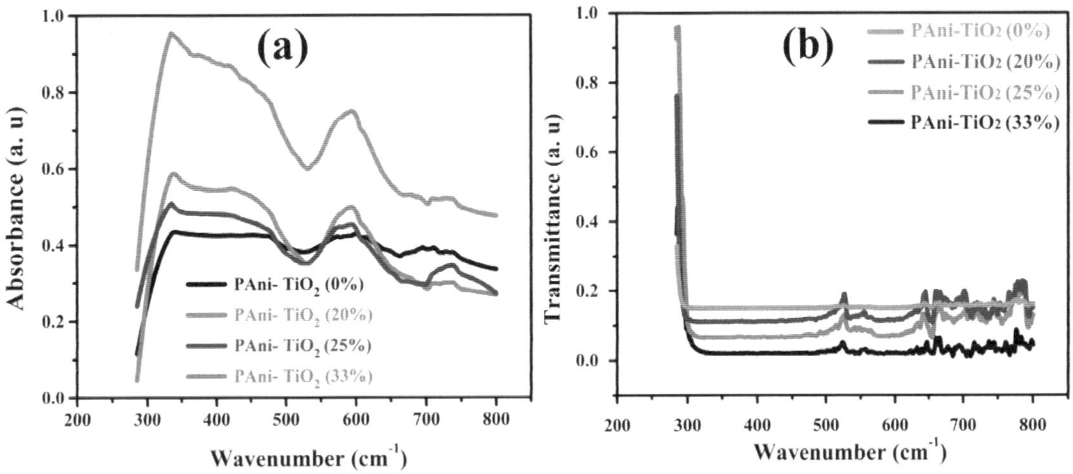

Figure 5: UV/Vis absorption an (a) transmittance (b) spectrum of PAni-TiO_2 nanocomposites at metal oxide ratios
(0, 20, 25, and 33)%.

The Tauc relationship was used to determine the samples' optical band gap (E_g) [49] . The Eg values vary depending on the structure and chemical composition, and they improve when the amount of TiO_2 powder added to PAni is increased (see Fig. 6). The greatest value of 2.26 eV was revealed by unmodified PAni, which was subsequently reduced by increasing the concentration ratios to reach the optimal value (1.61 eV) when composites was changed with 33% of nanopowders. According to previous studies, the absorption has been determined to depend on the short-range order in the amorphous state and defect state; the narrowing of the band gap in the current system may be related to the system's reduced disorder and increased density of defect states [50]. Therefore, it follows that the optical characteristics of active composites may be impacted by the addition of TiO_2 in various ratios to PAni structures.

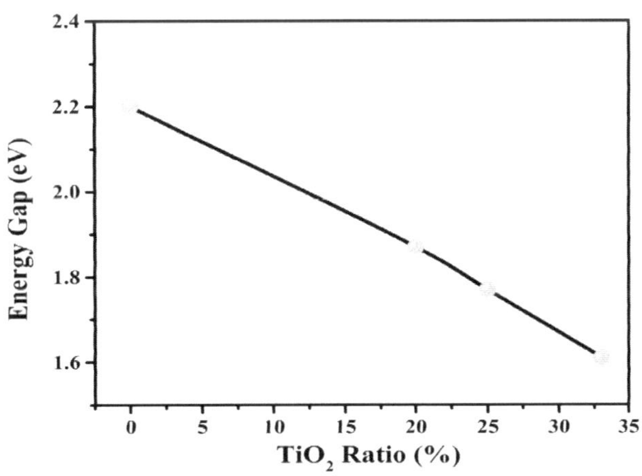

Figure 6: The value of E_g for PAni–TiO_2 composites as a function of ratio; TiO_2 NPs synthesized at a hydrothermal reaction time of 8h.

4.6. Electrical Properties

In light of this, it is possible to enhance the electrical properties of PAni by integrating TiO_2 NPs into the chains. [51]. Whereas, the oxygen ions are released and return to the surface of the TiO_2 grains as a result of the interaction between the nitrogen atoms in the PAni molecules and the O^- ions there. This traps electrons in the surface of the grains [52]. The electrical properties of the resultant composites were investigated using the 4PP (EQ-JX2008-LD) technique. The conductivity of the composites is shown in Table 3 for varied nanopowders concentrations. As can be shown from the resistivity investigations, the concentration ratios of TiO_2 NPs (0, 20, 25 and 33%) have a major effect on the conductivity of the films formed, causing the conductivity to rapidly improve to 4.77, 5.34, 6.04, and 6.58 S/m, respectively. The TiO_2 NPs' physicochemical interactions with the PAni chains, which result in a change in the oxidation state, are the main source of this activity, that has previously been explored by others [53], [54]. It is well known that PAni's electrical structure is necessarily impacted by the variation in oxidation level. As a result, it was reported that the polaron band for phase PAni in the composites was narrower than that for unmodified PAni (matching to absorption spectra), and it continuously became better by increasing the concentration of TiO_2 NPs [53]. Mobility of produced carriers (polarons) along PAni chains results from this in turn [54]. These exhibit greater values than those of earlier research by SU Bitao et al. ($2.86x10^{-7}$ S/m) [51]. and Xingwei Li et al. ($2.9x10^{-7}$ S/m) [55]. The flow of electrons from the surfaces of the polar O^{2-} TiO_2 NPs to the PAni chains, which results in their reduction, may be the cause of this notable increase in conductivity. Moreover, the creation of lattice or electronic defects, such as oxygen vacancies or interstitial Ti^{4+} ions, led to a comparable impact on modified PAni structures combined with a high concentration of TiO_2 nanopowders and was connected with an increase in n-type conductivity [56]. These findings demonstrate that a high TiO_2 NP concentration can enhance charge carriers, electronic interaction between PAni and TiO_2 composites, and conjugated length reduction in PAni chains, all of which promote electrical conductivity [57,58].

Table 3. Electrical performance for unmodified PAni, PAni-TiO_2 nanocomposites at various concentrations of TiO_2 NPs.

Sample	PAni-TiO_2 (0%)	PAni-TiO_2 (20%)	PAni-TiO_2 (25%)	PAni-TiO_2 (33%)
Conductivity (S/m)	4.77	5.34	6.04	6.58

4.7. Designed cells Performance

Using a normal set of parameters, such as 1.5 air mass (AM) and 25 °C, an I–V curves was determined using a solar tester device. As an acceptor, varying concentrations of TiO_2 NPs were combined with PAni's layers, which were created as a donor. The TiO_2 NPs' ability to transfer polarons and bipolarons over shorter distances along polymer chains contributes to an increase in conductivity in the PAni matrix when they are incorporated. A solar cell made of PAni-TiO_2 nanocomposites with various TiO_2 NP concentrations is shown in Fig. 7 with its J-V curves under light. Meanwhile, as seen in Table 4, the J-V behaviour used to determine PV characteristics of the solar cell device.

As can be seen, the J-V curve for the FTO/PAni-TiO_2 (33%)/Ag device is moved toward greater current densities than the one created with FTO/PAni-TiO_2 (0%)/Ag. In addition, the PAni-TiO_2 active layer created by including 33% of TiO_2 NPs into the PAni-structure gradually increases current density values in comparison to designs made with the pristine layer. Based on computed efficiencies, all of these produced devices perform better than other, comparable devices made with PAni-TiO_2 active layers [59 - 61].

One explanation for the greater current density might be:

- Incorporating a significant amount of TiO_2 NPs into the PAni matrix, which contributes to polarons and bipolarons, which transfer more quickly along PAni chains and boost the carriers transit in the PAni structure [62].

- The rapid increase in surface roughness decreases the reflected photons, therefore causes the generated carriers grow.
- A better donor/acceptor interface that facilitates simple exciton separation can result from an enhancement in the physicochemical interaction among p-type and n-type of PAni TiO_2 NPs, respectively [59].
- A decrease in the resistance at the interface between TiO_2 and PAni as a result of an increase in the PAni polarons number and the structural homogeneity of composites. [63].

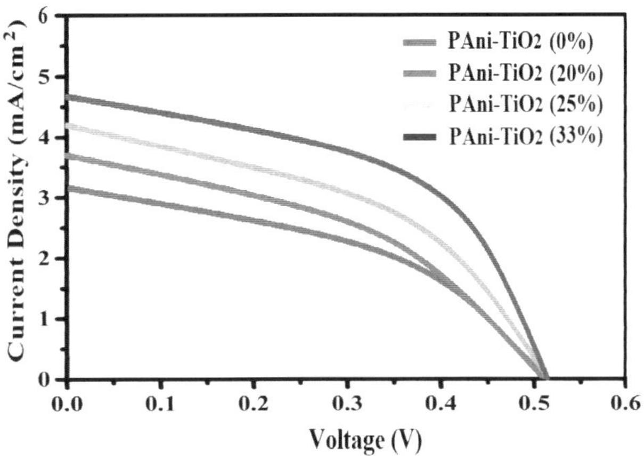

Figure 7. Characteristic J-V curves of the designed cells.

Table 4. Typical parameters for unmodified and modified PAni-TiO_2 composites cells.

Sample	V_{oc} (V)	J_{sc} (mA/cm^2)	FF (%)	Rs (Ω)	Rsh (kΩ)	η (%)
FTO/PAni-TiO_2 (0%)/Ag	0.51	3.11	20.8	65.7	1.43	0.33
FTO/PAni-TiO_2 (20%)/Ag	0.51	3.58	22.2	78.1	0.92	0.41
FTO/PAni-TiO_2 (25%)/Ag	0.51	4.17	28.8	46	1.28	0.61
FTO/PAni-TiO_2 (33%)/Ag	0.51	4.45	37.5	86.2	0.67	0.85

5. *Conclusion*

A chemical oxidative approach was effectively used to create a conductive PAni with an orthorhombic structural phase, which was then combined with various ratios of anatase TiO_2 NPs, including 0, 20, 25, and 33%. To create PAni-TiO_2 composites, which serve as the photoactive layer of organic solar cells, these varying proportions of nanopowders were incorporated during the polymerization of aniline monomer. Depending on the ratio of nanomaterial in composites, the evaluations of crystallite size, morphology, and roughness were significantly impacted. Whereas, the crystallite size decreased as the concentration of TiO_2 increased, going from 18.73 nm with 0% TiO_2 NPs to an optimum value of 12.89 nm at 33%. Meanwhile, the enhancement of the physicochemical interaction between the TiO_2 and PAni especially at higher concentrations led to an improvement in the nanoflakes' form and roughness characteristics. Additionally, following PAni modification with 33% of nanopowders, the optical band gap has an optimal value of 1.61 eV. As a result, the efficiency increased to 0.85% of the designed cell FTO/PAni-TiO_2(33%)/Ag as contrasted to 0.33% for the FTO/PAni-TiO_2(0%)/Ag cell. Other than the quick rise in surface roughness up to Ra = 2.61 m, which reduces the reflected photons and increases the generated charge carriers, structural factors are also responsible for the improvement in cell efficiency.

References

[1]. El-Hossary FM, Ghitas A, Abd El-Rahman AM, Ebnalwaled AA and Shahat, MA 2020 *IOP Conference Series: Materials Science and Engineering* **956** 012003.

[2]. Addonizio M L, Aronne A, and Imparato C 2020 *Applied Surface Science* **502** 144095.

[3]. El-Hossary FM, Ghitas A, Abd El-Rahman AM, Ebnalwaled AA, Fawey MH and Shahat MA 2020 *IOP Conference Series: Materials Science and Engineering* **762** 012001.

[4]. Hagfeldt A, and Graetzel M 1995 *Chemical reviews*, **95** (1) 49–68.

[5]. Najafi V, Ahmadi E, Ziaee F, Omidian H, Sedaghat H 2019 *J Polym Environ* **27** (4) 784–793.

[6]. Saeed M, Shakoor A, Ahmad E 2013 *Polym Sci Ser A* **55** (9) 549–555.

[7]. Kuwabata S, Takahashi N, Hirao S, Yoneyama H 1993 *Chem Mater* **5** (4) 437–441.

[8]. Gangopadyay R. Amitabha De 2000 *Chem Mater* **12** 608.

[9]. Song X L, Yan C Y, Huang S T, Zhang M W, Geng B Y, Meng R Bin 2013 *Advanced Materials Research* **750** 1098–1103.

[10]. Zagorny M, Bykov I, Melnyk A, Lobunets T, Zhygotsky A, Pozniy A, et al 2014 *J Chem Chem Eng* **8** (2) 118.

[11]. Parveen A, Roy AS 2013 *Adv Mater Lett* **4** (9) 696–701.

[12]. Diantoro M, Suprayogi T, Taufiq A, Fuad A and Mufti N 2019 *Materials Today: Proceedings* **17** 1197–1209.

[13]. Nemade K, Dudhe P and Tekade P 2018 *Solid State Sciences* **83** 99–106.

[14]. El-Hossary FM, Ghitas A, Abd El-Rahman AM, Ebnalwaled AA, Shahat MA and Fawey MH 2020 *IOP Conference Series: Materials Science and Engineering* **956** 012015.

[15]. Mostafaei A and Zolriasatein A 2012 *Progress in Natural Science: Materials International* **22** (4), 273–280.

[16]. Hanawalt J D, Rinn H W, Frevel L K 1938 *Ind Eng Chem Anal Ed* **10** (9) 457–512.

[17]. Selvan S T, Mani A, Athinarayanasamy K, Phani KLN, Pitchumani S 1995 *Mater Res Bull* **30** (6) 699–705.

[18]. Mani A, Athinarayanasamy K, Kamaraj P, Selvan S T, Ravichandran S, Phani KLNN, et al 1995 *J Mater Sci Lett* **14** (22) 1594–1596.

[19]. Monfared A H, Jamshidi M, Hashemi Monfared A, Jamshidi M, Monfared A H and Jamshidi M 2019 *Prog Org Coatings* **136** 105257.

[20]. Thieme K, Avramov I and Rüssel C 2016 *Scientific reports* **6** 25451.

[21]. Lai C, Li GR, Dou YY and Gao XP 2010 **55** (15) 4567–4572.

[22]. Ketch A, Burkhart M, Hwang T and Kim SS 2012 *Chemical engineering journal* **192** 262–268.

[23]. Hamann TW, Martinson AB, Elam JW, Pellin MJ and Hupp JT 2008 *Advanced Materials* **20** (8) 1560–1564.

[24] Shi L, Wang X, Lu L, Yang X, Wu X 2009 *Synth Met* **159** (23–24) 2525–2529.

[25] Kao S-Y, Lin Y-S, Chin K, Hu C-W, Leung M, Ho K-C 2014 *Sol energy Mater Sol cells* **125** 261–267.

[26] Zhang Q, Bao N, Wang X, Hu X, Miao X, Chaker M, et al 2016 *Sci Rep* **6** 38066.

[27] Seif A, Nikfarjam A, ghassem HH 2019 *Sensors Actuators B Chem* **298** 126906.

[28] Socrates G 2004 *John Wiley & Sons.*

[29] Rahman KH, Kar AK 2020 *J Environ Chem Eng* **8** (5) 104181.

[30] Monfared A H, Jamshidi M, Hashemi Monfared A, Jamshidi M, Monfared A H and Jamshidi M 2019 *Prog Org Coatings* **136** 105257.

[31] Quillard S, Louarn G, Lefrant S, MacDiarmid AG 1994 *Phys Rev B*, **50** (17), 12496
[32] Dobrucka R 2017 *Iran J Pharm Res IJPR* **16**(2) 756.
[33] Asha, Goyal SL, Kumar D, Kumar S, Kishore N 2014 *Indian J Pure Appl Phys* **52** (5) 341–347.
[34] Yang C, Dong W, Cui G, Zhao Y, Shi X, Xia X, et al 2017 *Acta* **247** 486–495.
[35] Zhang L, Liu P, Su Z 2006 *Polym Degrad Stab* **91** (9) 2213–2219.
[36] Min S, Wang F, Han Y 2007 *J Mater Sci* **42** (24) 9966–9972.
[37] Basavaraja C, Pierson R, Venkataraman A, Basavaraja S 2009 *Macromol Res* **17**(8) 609–615.
[38] Li X, Wang D, Luo Q, An J, Wang Y, Cheng G 2008 *J Chem Technol Biotechnol Int Res Process Environ Clean Technol* **83** (11) 1558–1564.
[39] Olad A, Behboudi S, Entezami AA 2012 *Bull Mater Sci* **35** (5) 801–809.
[40] Allen JE, Ray B, Khan MR, Yager KG, Alam MA, Black CT 2012 *Appl Phys Lett* **101** (6) 63105.
[41] Wang H, Han B, Lu J, Wu P, Cui W 2020 *Mater Lett* **260** 126906.
[42] Rahman KH, Kar AK 2020 *Mater Sci Semicond Process* **105** 104745.
[43] Jlassi K, Radwan AB, Sadasivuni KK, Mrlik M, Abdullah AM, Chehimi MM, et al 2018 *Sci Rep* **8** (1) 1–13.
[44] Deshpande NG, Gudage YG, Sharma R, Vyas JC, Kim JB, Lee YP 2009 *Sensors Actuators B Chem* **138** (1) 76–84.
[45] Xia H, Wang Q 2002 *Chem Mater* **14** (5) 2158–2165.
[46] Su S-J, Kuramoto N 2000 *Synth Met* **114** (2) 147–153.
[47] Cionti C, Pina C Della, Meroni D, Falletta E, Ardizzone S 2020 *Nanomaterials* **10** (3) 10030441.
[48] Takpire SR, Waghuley SA 2017 *J Energy Inst* **90** (1) 44–50.
[49] Pankove JI 1975 *Courier Corporation.*
[50] M. Reda S, M. Al-Ghannam S 2012 *Adv Mater Phys Chem* **02** (02) 75–81.
[51] Su B, Min S, She S, Tong Y, Bai J 2007 *Front Chem China* **2** (2) 123–126.
[52] Zagorny M, Bykov I 2014 *J Chem Eng Chem Res* **1** (1) 6–14.
[53] Pud AA, Ogurtsov NA, Noskov Y V, Mikhaylov SD, Piryatinski YP, Bliznyuk VN 2019 *Semicond Physics Quantum Electron & Optoelectron* **22** (4) 470–478.
[54] Ogurtsov NA, Noskov Y V, Fatyeyeva KY, Ilyin VG, Dudarenko G V, Pud AA 2013 *J Phys Chem B* **117** (17) 5306–5314.
[55] Li X, Chen W, Bian C, He J, Xu N, Xue G 2003 *Appl Surf Sci* **217** (1–4) 16–22.
[56] Deshpande NG, Gudage YG, Sharma R, Vyas JC, Kim JB, Lee YP 2009 *Sensors Actuators B Chem* **138** (1) 76–84.
[57] Gongadze E, Velikonja A, Perutkova Š, Kramar P, Maček-Lebar A, Kralj-Iglič V, et al 2014 *Electrochim Acta* **126** 42–60.
[58] Good RJ 1992 *J Adhes Sci Technol* **6** (12) 1269–1302.
[59] Geethalakshmi D, Muthukumarasamy N, Balasundaraprabhu R 2016 *Mater Sci Semicond Process* **51** 71–80.
[60] Ibrahim M, Bassil M, Demirci UB, Khoury T, Moussa GEH, El Tahchi M, et al 2012 *Mater Chem Phys* **133** (2–3) 1040–1049.
[61] Liu Z, Zhou J, Xue H, Shen L, Zang H, Chen W 2006 *Synth Met* **156** (9–10) 721–723.
[62] Diantoro M, Suprayogi T, Taufiq A, Fuad A, Mufti N 2019 *Mater Today Proc* **17** 1197–1209.
[63] Nemade K, Dudhe P, Tekade P 2018 *Solid State Sci* **83** 99–106.

ZnO Nanoparticles for Photocatalytic Methyl Orange Degradation: Hydrothermal Synthesis, Characterization, and Optimization

M. Abdelhamid Shahat [1,*], Ahmed Ghitas [1], F. M. El-Hossary [2], A. M. Abd El-Rahman [3,2]

[1] PV Unit, Solar and Space Research Department, National Research Institute of Astronomy and Geophysics (NRIAG), 11421 Helwan, Cairo, Egypt.
[2] Physics Department, Faculty of Science, Sohag University, 82524 Sohag, Egypt.
[3] King AbdulAziz University, Jeddah, KSA.
[4] Electronics & Nano Devices Lab, Physics Department, Faculty of Science, South Valley University, 83523 Qena, Egypt.

* Corresponding Author E. mail: m.abdelhamid999@gmail.com

Abstract. This study investigates the production of zinc oxide nanoparticles (ZnO NPs) utilizing an efficient and surfactant-free hydrothermal method for use as a photocatalytic agent in the purification of organic polluted water. The structural features, optical properties and photocatalytic performance of ZnO NPs have been investigated. XRD, SEM, HR-TEM, FTIR and UV–vis absorption spectroscopy were used as analytical and characterization techniques. The ZnO NPs prepared with hexagonal structure and interesting applied properties was obtained by adjusting the processing time at a low treatment temperature of 75 °C. The crystallite size and energy band gap are reaction time-dependent, which in turn affects the photocatalytic performance. ZnO NPs under UV irradiation show an efficiency of 93% in the photocatalytic decomposition of methyl orange (MO) dye. The value of ZnO NPs bandgap (3.38 eV) is suitable for allowing UV light absorption and rapid generation of charge carriers; creating photooxidation reactions for decomposing the organic dye.

Keywords. Hydrothermal, Nanoparticles, ZnO, Photodegradation, Water Purification.

1. Introduction

Globally, water contamination is treated as one of the critical and long-standing environmental problems. Water pollution is basically composed of organic and inorganic pollutants. Organic pollutants with non-biodegradable features have harmful toxicity effects to the environment and human life even with low concentrations. Accordingly, the treatment of organically polluted water is one of the research priority topics and still needs low cost and efficient sustainable solutions. Photocatalytic water treatment technology is usually considered in the decomposition of organic compounds into inorganics [1]. Metal oxide semiconductors with high chemical stability and good controllability are suitable materials for broad photocatalytic applications [2, 3]. They include solar fuel production [4, 5], water splitting [6, 7], photo-degradation and catalysis of other compound responses in polluted water [8, 9]. The TiO_2, ZnO, GO, rGO, WO_3, GaN and C_3N_4 are well recognized as binary metal oxide photocatalysts [10–14]. BiO_X (X = Cl, Br and I), Bi_2WO_6, $BiVO_4$, $ZnWO_4$, Ag_3PO_4 and Fe_2O_3 have been also used as ternary photocatalytic compounds [15–20].

Zinc oxide (ZnO) with nanostructural properties remains a capable nanomaterial used in the manufacture of PV cells industry and water treatment technology owing to its favorable circumstances of low cost and harmfulness, reliable photosensitive qualities with a good performance of photocatalytic [21–24]. ZnO's catalytic performance derives from its capacity to photo-generate electrons and holes, which function as powerful decreasing and oxidizing agents, facilitating the breakdown process of organic molecules.

Different techniques have been successfully employed in the preparation process of ZnO NPs. Precipitation [25–27], spray pyrolysis [28], thermal decomposition [29] and hydrothermal [30–33] are typical methods. In particular, the hydrothermal method is particularly applied to synthesize non-pure material with fine grain size [34, 35]. ZnO NPs can be prepared with different structures and morphologies including rods, wires, whiskers and plates. As practically known, the hydrothermal preparation method is controlled by adjusting different factors like the precursor reactants concentration, diversity in temperature and time [36,37]. Previous researchers successfully prepared ZnO NPs using hydrothermal technique at a treatment temperature of 120 °C for a reaction time of 24 h [38, 39].

This paper focuses on the prepare of ZnO NPs in a simplified and surfactant-free hydrothermal process for water treatment tenders. Furthermore, in order to evaluate their capability on the purification of organic-polluted wastewater, the catalytic efficiency of nanomaterials in the decomposition of methyl orange dye was investigated and the related results were discussed.

2. Experimental details

2.1. Preparation procedure

Fig. 1 shows the in-lab hydrothermal procedure of metal oxides nanoparticles. The ZnO NPs were produced by a hydrothermal reaction between Zinc nitrate hexahydrate ($Zn(NO_3)_2 .6H_2O$) and sodium hydroxide NaOH(s) at different reaction times. In a typical hydrothermal procedure, 3 g of Zinc nitrate hexahydrate was suspended in 20 ml of distilled water (DI). Further, a NaOH(s) was prepared by dissolving of 4.2 g of NaOH in 20 ml DI. The 5 mol/L solutions of Zinc nitrate hexahydrate and NaOH were mixed and continuously stirred for 2h at 950 rpm. The solution was sonicated for several minutes and transferred into an autoclave for heat-treatment at 75 °C for 8h. After that, the aqueous solution was permitted to cool down to RT for one day. The produced precipitates were washed using methanol and DI for several times, and finally dried at 110 °C for 3 h. These steps are repeated in the same preparation conditions (low hydrothermal temperature (75 °C)) and different times within the autoclave 12 and 18h.

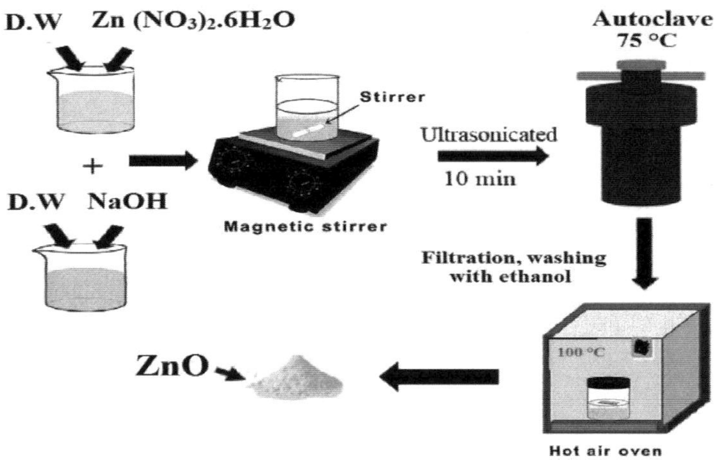

Figure 1. In-lab hydrothermal procedure of metal oxides nanoparticles

2.2. Analysis and Characterization

XRD was used to determine the phase composition of the ZnO NPs photocatalysts. A Philips diffractometer 1710 was used to produce a Cu K_α X-rays with a wavelength of λ=1.5418 Å. The morphological features of the synthesized photocatalysts were examined by HRSEM (JOUL SEM model JSM – 5500 - Japan) at an accelerating voltage of 10 KV. For obtaining a clear 2D imaging, a HR-TEM (JEOL, JEM- A 2100 – Japan) was used and operated at an accelerating voltage of 200 kV. Further, FTIR with KBr method (Jasco Model 4100_Japan) was used to detect the functional groups changes of the molecular structure of the produced materials at RT in the range of 4000- 400 cm^{-1} using a resolution of 4 cm^{-1}.

ZnO's optical characteristics were measured at RT using a computerized SPECORD 200 PLUS spectrophotometer. Transmittance and absorbance spectra were typically obtained at normal incidence within the wavelength ranging from 190 nm to 1100 nm. A standard sample preparation procedure is applied, in which ZnO nanopowders are dispersed in distilled water using ultrasonic, and put into a 1 cm quartz covet. The quantitative optical analysis calculates the direct energy band gap (E_g) using Tauc relationship [40, 41].

2.3. Photocatalytic performance

Additionally, by observing the photo-degradation of an aqueous solution of MO under UV irradiation, the catalytic performance of ZnO NPs is investigated. About 75 mg of the ZnO NPs were dissolved in 100 ml of MO solution (50 mg/L) using magnetic stirrer for several minutes in order to prepare 0.75 g/L of ZnO solution. Prior to photoirradiation using UV light, to achieve adsorption/desorption equilibrium, the mixture was stirred in a dark condition for 30 min. For photoactivating the ZnO NPs, the aqueous solution was stirred continuously under UV light illumination and after every 30 min a sample was taken for analysis. After that, the suspensions were gathered and centrifuged (3500 rpm, 17 min) to isolate the catalyst particles to determine the absorption peak of pure organic dye MO using UV–vis spectrophotometer.

3. Results and discussion

3.1. Structural investigation

Fig. 2 represents the XRD patterns of the ZnO photocatalysts synthesized at different reaction times of 8, 12 and 18 h. The crystalline structure and crystallite size are mainly identified. The ZnO photocatalysts indicated a hexagonal crystalline structure with corresponding diffraction peaks of (100), (002), (101), (102), (110), (103), (200), (112), (201), (004) and (202). These data are summarized in table 1 and provided a good agreement with the standard records of card No: 00-003-0888 [42, 43].

As previously approved, nano-size materials result in broad diffraction peaks in XRD patterns attributable to crystalline expansion [44]. Therefore, the XRD peak width is used in determining the crystallite size (D), and Crystal lattice strain broadening (ε). D and ε of the synthesized ZnO photocatalysts were simply calculated using a Williamson-Hall plot [45]. From the analysis of the ZnO NPs results, the broadening of the XRD peaks are decreased by increasing the reaction time; indicating a crystallinity enhancement of ZnO NPs. According to current interpretation, energetic nuclei may be generated at very short hydrothermal reaction durations, enhancing the nanoparticle growth rate and as a result, improving the crystallinity of ZnO NPs with larger crystallite sizes [45, 46].

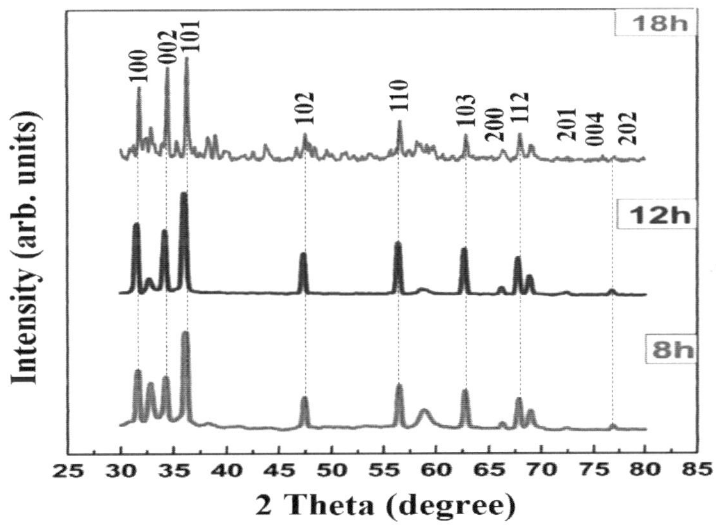

Figure 2. X-ray diffraction patterns of ZnO NPs prepared at various reaction times

Table 1. d_{hkl} for the synthesized ZnO NPs versus d_{hkl} for standard hexagonal ZnO crystals.

hkl	d_{hkl} (Å) Calculated			d_{hkl} (Å) for standard Hexagonal ZnO
	8 h	12 h	18 h	
100	2.82	2.84	2.81	2.81
002	2.61	2.62	2.60	2.60
101	2.48	2.50	2.48	2.48
102	1.91	1.92	1.91	1.91
110	1.63	1.63	1.63	1.62
103	1.48	1.48	1.48	1.48
200	1.41	1.41	1.41	1.41

112	1.38	1.38	1.38	1.38
201	1.62	1.36	1.36	1.36
004	1.30	1.30	-	1.30
202	1.24	1.24	1.35	1.24

Fig. 3 shows a relationship between Γcosθ and sinθ for samples synthesized at different reaction times. The values of D and lattice microstrain for the ZnO NPs at different reaction times of 8, 12 and 18 h are estimated and listed in Table 2. The average crystallite size for the prepared samples are found to be 8.32, 8.46 and 11.25 nm, while the corresponding values of microstrain are $35x10^{-3}$, 35 $x10^{-3}$ and 38 $x10^{-3}$, respectively. Khera, et al (2019) successfully synthesized ZnO NPs with different solvents, they found that crystallite size and microstrain are strongly influenced by various preparation conditions [42].

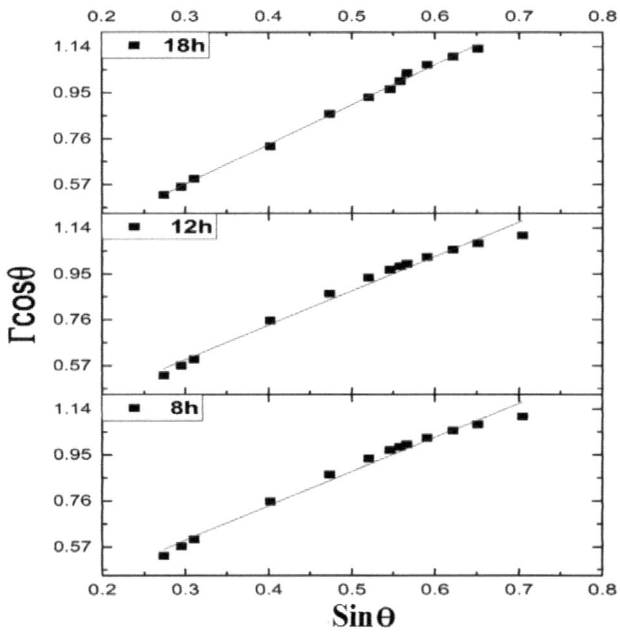

Figure 3. Plot of Γcosθ as a function sinθ for samples synthesized at different reaction times of 8, 12 and 18 h.

Table 2. Lattice parameters and crystallite size of the prepared materials.

Reaction time	Lattice-Parameters		Williamson-Hall	
	a (Å)	c (Å)	D size (nm)	Microstrain X (10^{-3})
8 h	2.83	4.42	8.32	38
12 h	2.82	4.41	8.46	37
18 h	2.57	4.25	11.25	34

3.2. FTIR spectra

Fig. 4 displays the FTIR absorption spectra of ZnO NPs samples produced at different reaction times of 8, 12 and 18h. The FTIR spectra of all samples show two absorption bands at peak position of 2419 with limited shifted and 3483 cm^{-1}. They found to be related to stretching vibration bonds of C=O and O–H, respectively. Both bonds are existed between the adsorbed water molecules and indicated higher amount of hydroxyl group [48]. Further, an absorption peak is detected between 450 and 500 cm^{-1} and indicated a Zn–O group that found in a stretching vibration mode. This is in agreement with previous research work [48, 49]. Furthermore, when the hydrothermal reaction duration increases, a small peak shift is seen at wavenumber 500 cm^{-1}. This may be attributed to changes in the crystalline lattice and the morphology of the prepared metal oxide [42].

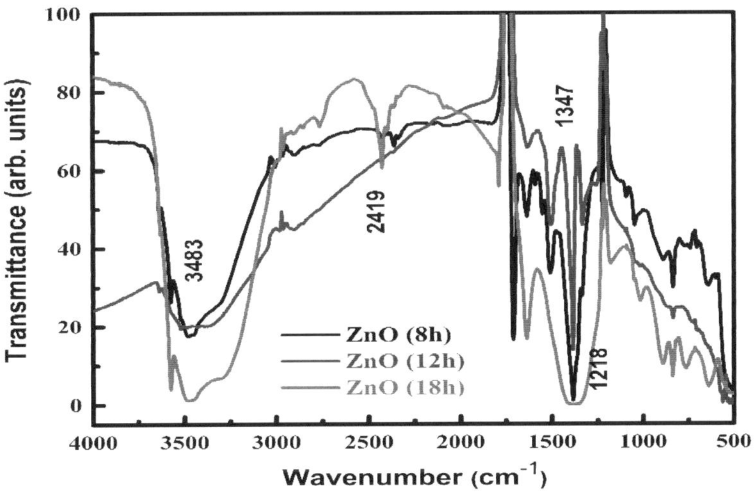

Figure 4. FTIR spectra of the synthesized samples.

3.3. *Morphology*

SEM and HR-TEM were used to investigate the shape and size dependency of ZnO NPs on hydrothermal treatment time, as shown in Fig. 5. As observed, the morphology of ZnO NPs synthesized at low hydrothermal treatment time is characterized by partially uniform particle shape and distributed particle size. On the other side, the morphology of ZnO NPs showed a structural change shape with an increase in particle size with the increase of hydrothermal treatment time from 8 to 18 h. Previously, photocatalytic performance is known to be influenced by the concentration of photo-generated charge carriers on the metal oxide surface. The concentration of charge carriers is directly correlated with the surface-to-volume ratio of the nanoparticles which in turn depends on preparation condition [50–53].

Figure 5. (a, b, c) SEM and (d, e, f) HRTEM images for synthesized ZnO NPs at various preparation times (8, 12, 18h), respectively.

3.4. Optical behavior

Fig. 6a,b illustrates absorption and transmittance spectra obtained within UV-Vis range for the as-prepared samples. One can observe a decrease in the light absorption by increasing the treatment time, while the transmittance values increased. In the absorption spectrum, a shift in the absorption edge towards shorter wavelengths was observed. It is shifted from 260 nm to 300 nm when the reaction time increased from 8 to 18 h. This is greatly lower value compared to that reported for bulk ZnO (380 nm) [55].

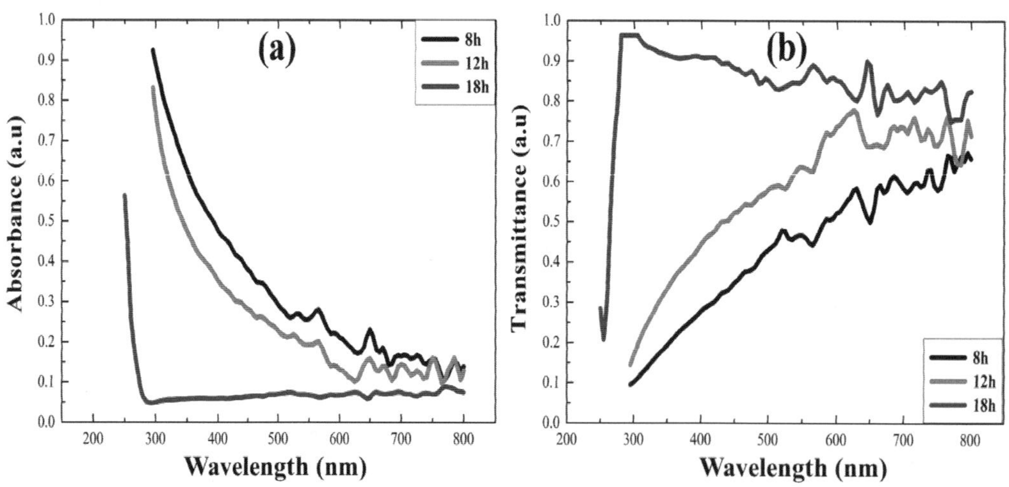

Figure 6a,b. Absorbance and transmittance spectra obtained within UV-Vis range for prepared ZnO NPs.

IOP Conf. Series: Materials Science and Engineering 1269 (2022) 012010 doi:10.1088/1757-899X/1269/1/012010

Fig. 7 shows a variation of E_g as a function of hydrothermal reaction time. It is noticed that the Eg increases as the reaction time increases. Further, the calculated E_g values are 3.38, 3.67 and 4.63 eV corresponding to samples synthesized at various preparation times of 8, 12 and 18 h, respectively. Other researchers succeeded to synthesize ZnO NPs hydrothermally with relatively higher E_g of 3.62 eV and 4.9 eV [55, 56]. The relatively changes in E_g value is mainly owing to the variation in the localized state density in E_g [57, 58]. Further, synthesizing ZnO with low crystallite size can be considered as contributing factor [60].

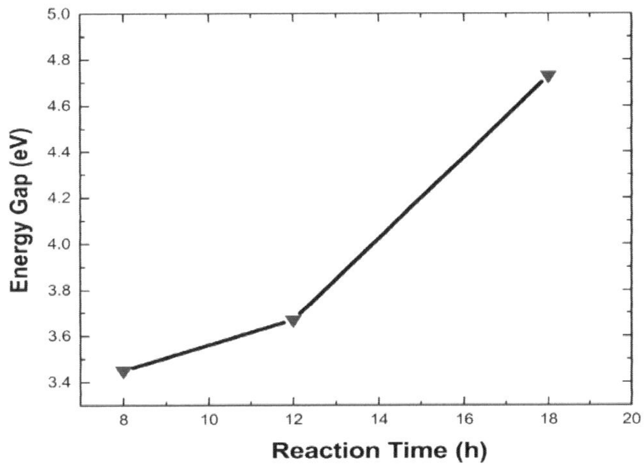

Figure 7. Variation of energy band gap (E_g) of ZnO NPs as a function of hydrothermal reaction time

3.5. Photocatalytic performance

In the presence of a photocatalytic activator composed of metal oxide NPs, several organic molecules are effectively broken down. The decomposition procedure is performed by photooxidation reaction. The e^-/ h^+ pairs are created from the metal oxide NPs surface after illumination by UV light. The lower energy gap of metal oxide NPs leads to a speedy creation of e^-/ h^+ pair. Then, the e^-/ h^+ pairs migrated to the NPS surface and strongly react with the $O^{\bullet}_2$ and $O\bar{H}^{\bullet}$ complicated in the MO suspensions. These reactions lead to radical's creation of hydroxyl (-OH), superoxide anions ($O\bullet^{-2}$) and hydroperoxyl ($OOH^{\bullet}$). The most potent oxidizing species known to destroy harmful organic contaminants are hydroxyl radicals, which create nontoxic molecules of gaseous CO_2 and H_2O [61].

ZnO photocatalytic activity is estimated by determining the concentration MO dye as a standard test material using light absorbance phenomena. Further, in the region of wavelengths from 300 to 600 nm, the MO-ZnO suspensions was examined using the UV-Vis Spectrophotometer. Meanwhile, the Beer-Lambert law is used to compute the MO concentration. For accurate comparison, calculation on suspension with zero percent of photocatalytic material was applied.

Fig. 8 illustrate absorption spectra for a mixed ZnO NPs/MO suspension obtained after different time intervals for nanomaterials synthesized at various treatment times of 8, 12 and 18h. A maximum absorption peak is noticed about 464 nm for all investigated samples. MO's absorption peak intensity was shown to slowly decrease as UV irradiation time increased. This exhibits a gradual photodegradation for the MO organic dye which is nearly decomposed after relatively long exposure time (270 min).

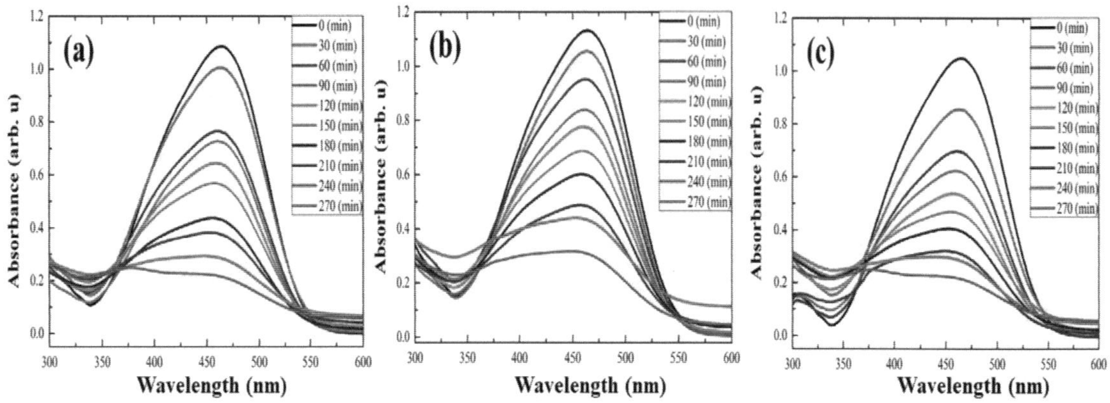

Figure 8. Absorption response for a mixed ZnO/MO suspension obtained after different time intervals mixed for synthesized samples.

Figure 9 depicts a typical colors matching experiment done using MO dye combined with ZnO-8h. An orange color gradient has observed with an increase of the UV irradiation time. In general, a maximum photocatalytic performance is obtained using a photocatalytic ZnO NPs prepared at relatively shorter treatment time of 8h. This is clearly observed from the faster degradation of the MO dye in MO-ZnO NPs suspension mixed with ZnO-8h. The use of ZnO NPs produced at a relatively shorter treatment process (8h) can significantly affect the properties of surface defects that decrease the recombination rate of photogenerated free carriers (e-h pairs); resulting in improved the photocatalytic efficiency of ZnO.

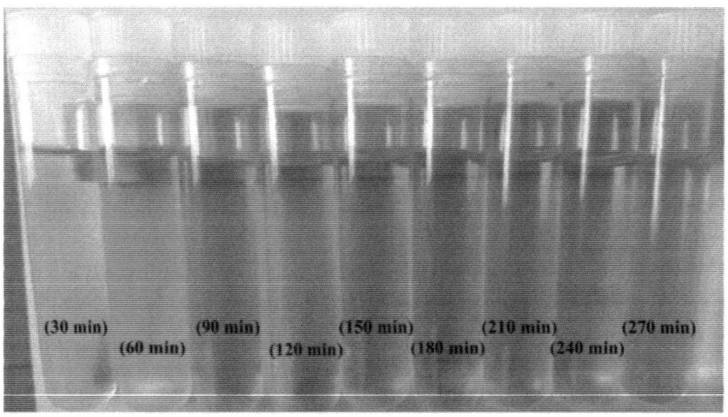

Figure 9. Typical colors matching test using a mixture of MO dye and ZnO-8h.

On another way, a dimensionless ratio between the initial concentration of MO dye upon equilibrium adsorption (C_0) and its concentration during the reaction process (C) is the simplest definition of the photodegradation efficiency. As previously investigated, it is difficult to decompose MO without using a photocatalytic agent irradiated with the presence of UV light source of or sunlight [62]. The kinetic rate constant (K) is calculated from the following equations [63].

$$-\frac{dCo}{dt} = KC_0 \quad (1)$$

$$\ln\left(\frac{C0}{C}\right) = Kt \quad (2)$$

Where C_0 (mg/L) is the initial dye concentration, besides, the ultimate dye concentration is represented by C (mg/L), and time is represented by t (min). In the presence of photocatalytic ZnO NPs, the MO photo-degradation under UV light is shown in kinetic form in Fig. 10, including concentration ratio (C/C_0) and Pseudo-first order Kinetics ln (C_0/C) as a function of irradiation time. Both of them demonstrate the same scientific meaning that is characterized by a linear relationship. The pseudo first order kinetics is gradually increased as the irradiation time increased. The photodegradation rate (K) of MO dye is calculated and shown in Fig 11 as a function of hydrothermal treatment time of ZnO NPs. It is reduced from 0.00813, 0.007 and 0.005 min^{-1} with an increase of hydrothermal reaction time from 8 to 12 and 18 h, respectively.

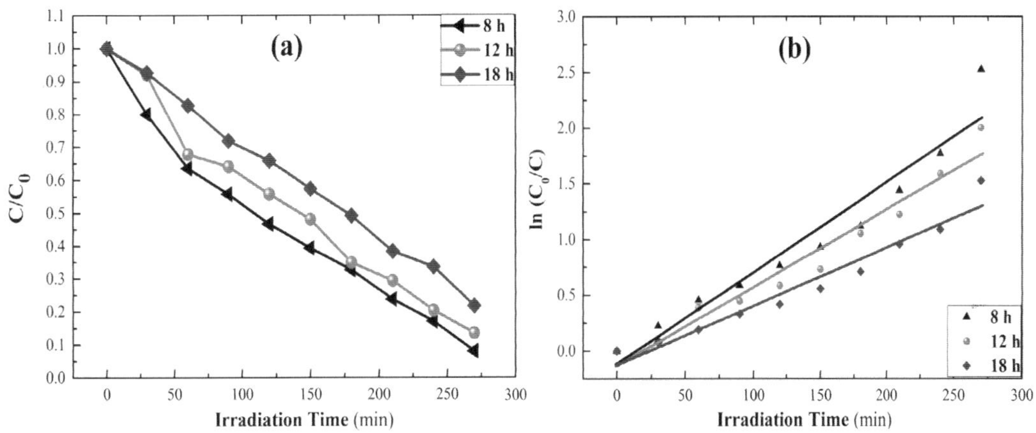

Figure 10. Kinetic of the MO degradation under UV irradiation in the existence of photocatalytic ZnO NPs: a) concentration ratio (C/C_0) and b) Pseudo-first order Kinetics ln (C_0/C) as a function of irradiation time.

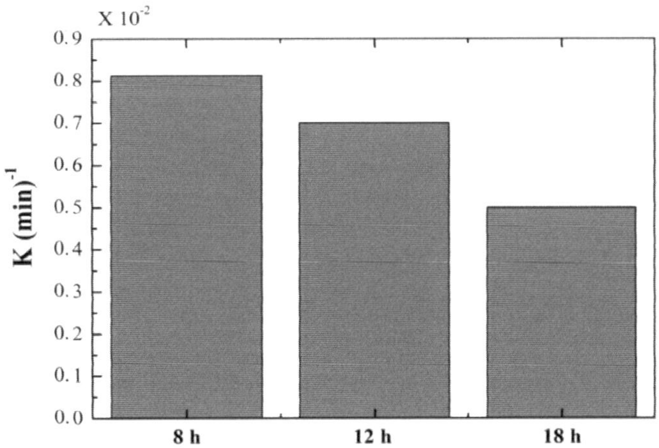

Figure 11. Variation of photodegradation rate of the MO dye as a function of hydrothermal treatment time of ZnO NPs.

Fig. 12 illustrates the photodegradation efficiency (η %) via irradiation time obtained for the photocatalytic ZnO synthesized at various hydrothermal reaction times. It has been found that the sample prepared at reaction time of 8h exhibits highest photocatalytic efficiency (93.6 %) compared to other prepared samples (86.3% and 78.3% for 12 and 18h, respectively). This is mainly due to the impact of structural and optical properties [64,65]. The prepared ZnO NPs at 8 h has relatively lower D of 8.32 nm and narrower Eg of

3.38 eV compared to that of other samples. This will be associated with the higher generation rate of e/h pairs which are trapped with metal ions.

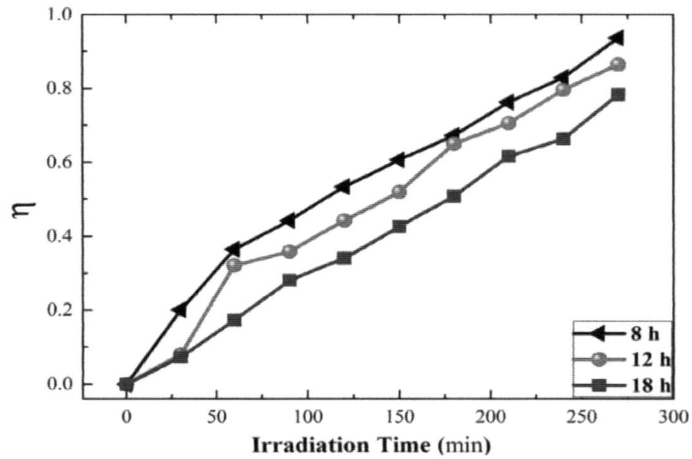

Figure 12. Photodegradation efficiency (η %) obtained for synthesized nanopowders as a function of UV-irradiation time.

4. Conclusion

ZnO NPs were effectively produced at a treatment temperature of 75 °C to serve as photocatalysts in the purification of organic wastes. Adjusting the treatment duration has a significant impact on the crystallite size and energy gap values of ZnO NPs. In details, an effective decrease in the E_g values from 4.63 to 3.38 eV accompanied with a decrease in the crystallite size from 11.25 to 8.32 nm, when the treatment time decreased from 18 to 8h. Additionally, it has been proposed that the lower band gap of ZnO NPs with surface defects expected to be formed on the nanoparticles has resulted in a high photogeneration rate of e^-/h^+ pairs, which in turn has been considered to improve the photodegradation efficiency. Moreover, the results showed a high photocatalysis efficiency of 93% of ZnO NPs in the chemical degradation of the MO after exposure time of 270 min under UV irradiation. This supports the use of ZnO NPs as an efficient photocatalytic agent in wastewater purification.

References

[1]. Dong S, Feng J, Fan M 2015 *Rsc Adv* **5** (19) 14610-14630
[2]. Wang W, Tadé MO, Shao Z 2015 *Chem Soc Rev* **44** (15) 5371-5408
[3]. Chen X, Liu L, Peter YY, Mao SS 2011 *Science* **331**(6018) 746-750
[4]. Sun K, Shen S, Liang Y, Burrows PE, Mao SS, Wang D 2014 *Chem Rev* **114** (17) 8662-8719
[5]. Yu J, Jin J, Cheng B, Jaroniec M 2014 *J Mater Chem A* **2** (10) 3407-3416
[6]. Boukhvalov DW, Son Y-W, Ruoff RS 2014 *ACS Catal* **4** (6) 2016-2021
[7]. Hisatomi T, Kubota J, Domen K 2014 *Chem Soc Rev* **43** (22) 7520-7535
[8]. Peng X, Fu J, Zhang X, et al 2013 *Nanosci Nanotechnol Lett* **5** (12) 1251-1257
[9]. Wu H-S, Sun L-D, Zhou H-P, Yan C-H 2012 *Nanoscale* **4** (10) 3242-3247
[10]. Wang S, Pan L, Song J-J, et al 2015 *J Am Chem Soc* **137** (8) 2975-2983
[11]. Liu J, Hu Z-Y, Peng Y, et al 2016 *Appl Catal B Environ* **181** 138-145
[12]. Liu J, Margeat O, Dachraoui W, Liu X, Fahlman M, Ackermann J 2014 *Adv Funct Mater* **24** (38)

6029-6037
[13]. El-Hossary FM, Ghitas A, El-Rahman AMA, Shahat MA, Fawey MH 2021 *Vacuum* 110158
[14]. El-Hossary FM, Ghitas A, Abd El-Rahman AM, Ebnalwaled AA, Shahat MA, Fawey MH 2021 *J Mater Sci Mater Electron* 1–14
[15]. Cheng H, Huang B, Dai Y 2014 *Nanoscale* **6** (4) 2009-2026
[16]. Li C, Chen G, Sun J, et al 2015 *ACS Appl Mater Interfaces* **7** (46) 25716-25724
[17]. Saison T, Chemin N, Chanéac C, et al 2015 *J Phys Chem C* **119** (23) 12967-12977
[18]. Bai X, Wang L, Zhu Y 2012 *ACS Catal* **2** (12) 2769-2778
[19]. Martin DJ, Liu G, Moniz SJA, et al 2015 *Chem Soc Rev* **44** (21) 7808-7828
[20]. Miseki Y, Majima Y, Gunji T, Sayama K 2014 *Chem Lett* **43** (10) 1560-1562
[21]. Li L, Liu X, Zhang Y, et al 2013 *ACS Appl Mater Interfaces* **5** (11) 5064-5071
[22]. Nagaveni K, Sivalingam G, Hegde MS, Madras G 2004 *Environ Sci Technol* **38** (5) 1600-1604
[23]. Linsebigler AL, Lu G, Yates Jr JT 1995 *Chem Rev* **95** (3) 735-758
[24]. Achouri F, Corbel S, Aboulaich A, et al 2014 *J Phys Chem Solids* **75** (10) 1081-1087
[25]. Öner M, Norwig J, Meyer WH, Wegner G 1998 *Chem Mater* **10** (2) 460-463
[26]. Costa ME V, Baptista JL 1993 *J Eur Ceram Soc* **11** (4) 275-281
[27]. Trindade T, de Jesus JDP, O'Brien P 1994 *J Mater Chem* **4** (10) 1611-1617
[28]. Liu T-Q, Sakurai O, Mizutani N, Kato M 1986 *J Mater Sci* **21** (10) 3698-3702
[29]. Andres-Verges M, Martinez-Gallego M 1992 *J Mater Sci* **27** (14) 3756-3762
[30]. Chen Q, Qian Y, Chen Z, Zhou G, Zhang Y 1995 *Mater Lett* **22** (1-2) 93-95
[31]. Andeen D, Loeffler L, Padture N, Lange FF 2003 *J Cryst Growth* **259** (1-2) 103-109
[32]. Sue K, Kimura K, Yamamoto M, Arai K 2004 *Mater Lett* **58** (26) 3350-3352
[33]. Sue K, Kimura K, Arai K 2004 *Mater Lett* **58** (25) 3229-3231
[34]. Ohshima E, Ogino H, Niikura I, et al 2004 *J Cryst Growth* **260** (1-2) 166-170
[35]. Yoshimura M 1998 *J Mater Res* **13** (4) 796-802
[36]. Mclaren A, Valdes-Solis T, Li G, Tsang SC 2009 *J Am Chem Soc* **131** (35) 12540-12541
[37]. Flores NM, Pal U, Galeazzi R, Sandoval A 2014 *RSC Adv* **4** (77) 41099-41110
[38]. Lai Y, Meng M, Yu Y, Wang X, Ding T 2011 *Appl Catal B Environ* **105** (3-4) 335-345
[39]. Levkevich EA, Yukhnovets O, Moshnikov VA, Maximov AI 2019 *IEEE Conference of Russian Young Researchers in Electrical and Electronic Engineering (EIConRus)* 777-779
[40]. Pankove JI, 1975 Courier Corporation
[41]. Fox M 2002 Published online
[42]. Khera S, Chand P, 2019 *Chinese J Phys* **57** 28-46
[43]. El-Hossary F M, Ghitas A, Abd El-Rahman A M, Ebnalwaled A A, Fawey M H, Shahat M A 2020 *IOP Conference Series: Materials Science and Engineering* **762** 012001
[44]. Giannini C, Ladisa M, Altamura D, Siliqi D, Sibillano T, De Caro L 2016 *Crystals* **6** (8) 87
[45]. Hamann TW, Martinson ABF, Elam JW, Pellin MJ, Hupp JT 2008 *Adv Mater* **20** (8) 1560-1564
[46]. Davar F, Mohammadikish M, Loghman-Estarki MR, Masteri-Farahani M 2014 *Ceram Int* **40** (6) 8143-8148
[47]. Yadav SK, Jeevanandam P 2015 *Opt Mater (Amst)* **46** 209-215
[48]. Rath PP, Behera SS, Priyadarshini B, et al 2019 *Appl Surf Sci* **491** 256-266
[49]. Teow S-Y, Wong MM-T, Yap H-Y, Peh S-C, Shameli K 2018 *Molecules* **23** (6) 1366
[50]. Guo Y, Cao X, Lan X, Zhao C, Xue X, Song Y 2008 *J Phys Chem C* **112** (24) 8832-8838
[51]. Bhatia S, Verma N 2017 *Mater Res Bull* **95** 468-476
[52]. Byzynski G, Pereira AP, Volanti DP, Ribeiro C, Longo E 2018 *J Photochem Photobiol A Chem* **353** 358-367
[53]. Manoharan G, Karuppiah M, Sahu NK, Hembram K 2018 *Ceram Int* **44** (7) 7837-7843
[54]. Akir S, Barras A, Coffinier Y, Bououdina M, Boukherroub R, Omrani AD 2016 *Ceram Int* **42** (8) 10259-10265
[55]. El-Hossary FM, Ghitas A, Abd El-Rahman AM, Ebnalwaled AA, Shahat MA 2020 *IOP Conference Series: Materials Science and Engineering* **956** 012003
[56]. Wasly HS, El-Sadek MSA, Henini M 2018 *Appl Phys A* **124** (1) 76
[57]. Preethi S, Anitha A, Arulmozhi M 2016 *Indian J Sci Technol* **9** (40) 1-6
[58]. Muhammad FF, Aziz SB, Hussein SA 2015 *J Mater Sci Mater Electron* **26** (1) 521-529
[59]. Rozra J, Saini I, Sharma A, et al 2012 *Mater Chem Phys* **134** (2-3) 1121-1126
[60]. Kramer IJ, Sargent EH 2014 *Chem Rev* **114** (1) 863-882

[61]. Ahmadi M, Ghasemi MR, Rafsanjani HH 2011 *J Mater Sci Eng* **5** (1) 87
[62]. Hu J, You N, Yu Z, Zhou G, Xu X 2016 *J Appl Phys* **120** (7) 74301
[63]. Singh N, Chakraborty R, Gupta RK 2018 *J Environ Chem Eng* **6** (1) 459-467
[64]. Zayed M, Samy S, Shaban M, Altowyan AS, Hamdy H, Ahmed, AM 2022 *Nanomaterials* **12**(6) 989
[65]. Mohamed F, Hassaballa S, Shaban M, Ahmed AM 2022 *Nanomaterials* **12**(2) 235

Impact of atmospheric extinction coefficient on the solar spectrum at selective wavelengths at Helwan in Egypt

Heba Zenhom[1,*], Ahmed Ghitas[1], Mohamed M. Abdelwahab[2], Mohamed S. Alnawawy[2], Amir H. Hassan[1], Ahmed Abulwfa[1]

[1]Solar and Space Research Department, National Research Institute of Astronomy and Geophysics (NRIAG), Helwan, Egypt

[2]Astronomy, Space Sciences and Meteorology Department, Faculty of Science, Cairo University, Giza, Egypt

***Corresponding author: hebazenhomhassan@gmail.com**

Abstract. Solar irradiance alterations due to the various atmospheric conditions have received great scientific concern. These alternations can be investigated by both monochromatic extinction coefficient and atmospheric transparency. In this work, continuous measurements of solar spectral irradiation using the Spectroradiometer for six selective wavelengths (350, 368, 500, 615, 780, and 870 nm) are carried out at the Helwan station, (NRIAG) in Egypt. Twelve clear-sky days of continuous observations have been monitored under different atmospheric conditions. The results have shown that using Langley method under cloudless sky for the total atmospheric extinction at the selective wavelengths was correlated. The extinction coefficient at the selected wavelengths shows obvious diurnal variations in each season. The perceptible effect of local climate has been obtained from the comparison of seasonal extinctions. The highest value of extinction coefficient was 0.59 km^{-1} at 615 nm in summer while the lowest one at 350 nm was 0.07 km^{-1} in winter. Moreover, it has been found that the afternoon values of transparency have a shift toward the longer wavelengths.

Keywords: Extinction coefficient, Atmospheric transparency, Spectroradiometer, Langley method.

1. Introduction

Solar radiation attenuation caused by atmospheric constituents and aerosols in the spectral regions of ultraviolet, visible and near-infrared can provide important information about the spectral variation [1]. By estimating the total extinction spectrum, characteristics of suspended particles can be obtained in air masses of various origins [2]. As greater cloudiness causes more attenuation process, hence the solar radiation attenuation can be quantified by atmospheric transparency parameter. [3]. The transparency of an atmospheric environment (through the lower atmosphere) can be estimated from its optical depth. [4].

The atmospheric extinction has been studied by Cerqueira et al. in Brazil for two selected wavelengths 500 nm and 670 nm via sun photometer [5]. A relationship between the horizontal extinction coefficient and particulate matter PM_{10} has been evaluated by Che. et al. in the period from 1980 to 2002 in Xi'an City [6]. They found that there are similar seasonal variations between the horizontal extinction coefficient and the concentration of PM_{10}, which depicts the effect of PM_{10} on the drop in visibility site. The atmospheric extinction coefficient variability has been investigated by other authors such as Cerqueira et al. and Aljawi.et al. on the basis of sun photometer measurements

[5, 7]. the reported investigations have been relied on Langley method using visible radiation bands of 400, 500, 670, 700 and 870 nm.

With the progression of large solar tower projects, it has been found that the study of atmospheric extinction is important for experimental and modelling levels. In 2017 Hanrieder et al. found that high extinction coefficient values were resulted from both high ambient $PM_{2.5}$ concentrations (68.5±21.7 μg/m^3) and high relative humidity (69.7±8.6%) [8]. In 2018 Trashorras et al. worked on the analyzing of the attenuation processes of the solar radiation considering the effects of both climate and meteorological conditions [3]. Khalil et al. found that the highest values of extinction coefficient in (GIR) solar radiation and the lowest values occur in (GUV) solar radiation, while the values of (G) solar radiation occur between them [9]. Generally, it can be concluded that during the partial eclipse there is an increase in extinction coefficient values; meanwhile the extinction coefficient has minimum values at noon time which could be attributed to the less value of air mass at noon. In this paper, we present the atmospheric extinction effects on the spectral variations. The used data is based on outdoor measurements on cloud free days.

2. Site Observation and Measurements

In this work, the effects of the mass extinction cross section on the spectral intensity of incident solar radiation has been studied. The solar spectral variation versus air masses are carried out by using a Spectro-radiometer fixed on a hilltop of the new building of the National Research Institute of Astronomy and Geophysics (NRIAG) at Helwan, (Lat. 29. 866°N and Long. 31. 20°E and 130 m above sea level) Egypt.

2.1. Spectral intensity measurements

The measurements on the site are continous from 2011 until now and for the present study data were collected for 2011 with a fiber optic Spectroradiometer (USB650). This Spectroradiometer is a spectrally automated instrument for measuring solar radiation; the spectra were recorded in continuity from local sunrise to local sunset. The bandwidth of the instrument is 350– 1000 nm and the Spectroradiometer spectral sensitivity of the charge coupled device (CCD) is 75 photons per count at 400 nm. A base mounting and glass dome covering have been designed for the protection of Spectroradiometer sensor from the environmental effects, and it has been calibrated for field testing. For measuring the irradiance of the spectrum at different angles of incidence, there is a cosine correction head is used for the Spectroradiometer sensor. Each recorded spectrum covers the spectral range from 350 nm to 1000 nm. The spectral measurements are programmed to be performed for every 1 min and then stored. The analyses have been made form sunrise up to sun set at selective wavelengths 350, 368, 500, 615, 780, and 870 nm. The effects of air mass have been taken on cloudless days; where we have measurements of 12 clear days, 3 days in each season of the year.

2.2 Meteorological conditions

The climate of Helwan is subtropical and consists of two traditional seasons (summer and winter) and two transition seasons (spring and autumn) generally, the topography around the site is hot, dry and dusty. Furthermore, Helwan is an industrialized Town suffering from the presence of strong pollution sources in the West such as (a power Station and Port Land Cement Company), Southwest. Hence, the conditions of observation at this site are influenced by these pollution sources. Approximately 50% of the air pollution of the site is from cement factories influenced by the prevailing wind direction from the north and northeast. This wind represents about 50% of the total directions.[6].

3. Methodology

According to Bouguer's law, the solar radiation intensity depleted by the earth's atmosphere can be expressed as in equation 1:

$$I_{n\lambda}=I_{0n\lambda}\exp(-\alpha\, m) \qquad (1)$$

Where $I_{0n\lambda}$ is the spectral extraterrestrial irradiance at the mean sun-earth distance, α is the monochromatic extinction or attenuation coefficient, m is the optical path length. This law applies only for a monochromatic radiation [10]. The optical air mass (m) can be defined as the ratio of the optical path through the atmosphere to the vertical path in the zenith direction, where all elements (air molecules and aerosols) deplete incomes of solar radiation along the atmosphere till reaching earth. Using of Langley method, derived from lambert's Law, we can find the atmospheric extinction coefficient variability [11, 12]. Using Eq. (1), Langley method can be expressed in the linearization form by putting it in a logarithmic formula, and then extract the fitted line equation y = b - ax , as the following: $\ln I_{n\lambda}= \ln I_{0n\lambda} - \alpha\, m$, Where $x = m$, $y= \ln I_{n\lambda}$, $b= \ln I_{0n\lambda}$, $a = \alpha_{total}$
x and y are variables, b is the extraterrestrial constant for each wavelength, and a is the total extinction coefficient (α_{total}). The total atmospheric extinction coefficient (α_{total}) is the sum of extinction coefficients created by scattering of air molecules (α_r) and absorption of gases (α_g) like ozone (α_{O3}) and aerosols (α_a). Hence, by drawing a relationship of m and Ln I_λ we can get the values of total extinction coefficient (α) from the negative slope of the straight-line from linear regression via the least squares method according to the next equation 2:

$$\alpha = -\ln (I_\lambda \;/\; I_{0\lambda}) \; (1/m) \qquad (2)$$

An important parameter has been also investigated which is air transparency (T_r) and given by

$$T_r= e^{-\alpha} \qquad (3)$$

Where α is the total extinction coefficient. Known of the day time, site's latitude and longitude, the air mass can be valued by simple geometrical considerations resulting in non-refractive plane parallel atmosphere formulation of uniform density.

$$m = \sec\theta \qquad (4)$$

Where m is the optical air mass and θ is the sun's zenith angle [5,9].

4. Results and Discussion

From morning to near the solar noon the solar spectra measurements through a distinctive clear sky day on 17 September 2011 is presented in Figure 1. The organized global spectrum in different air mass have been overlapped for contrast. Peaks of absorption by ozone, oxygen, and water molecules in the measured spectra can be perceived. During time of the day, changed parts of the spectrum vary in a different way. It is evident that the most important deviations are associated with the band (380–780 nm) represented by the visible (VIS) part of the spectrum in contrast to the deviations in the ultraviolet (UV) and the infrared (IR) regions. It appears that the absorption peaks occurred in the spectra are owing to the solar Fraunhofer absorption lines and take place at frequent wavelengths in each spectra. Air molecules and dust absorb and scatter light; consequently they are considered the main components in the atmosphere responsible for the energy decreasing from solar radiation. At noon, sunlight looks as if white which can be attributed to these atmospheric components experiences a comparatively regular decrease through the visible spectrum. Nevertheless, light undergoes extra efficient absorption and scattering in case of short wavelength of high energy as it passes longer through the atmosphere. Thus, the sun seems considerable redder with lesser intensity than for central of the day at the sunrise and sunset.

4.1. **Air** *mass impact on solar spectrum*

In this section, the characteristics of atmospheric attenuation for Helwan site have been described. Where, the data collected through the periods of measurement before and afternoon are fitted to a linear function and presented in Langley plot. The Figures from 2 to 4 show the plotting obtained with respective fitted lines for bands in UV, visible and near IR between the air mass and ln of the spectral intensity of incident solar radiation at six monochromatic wavelengths (350, 368, 500, 615, 780 and 870 nm). They are statistically analyzed on selected twelve clear days covering the four seasons of the year before and afternoon, where each set of three days represents a particular season. That means, 21 January was selected for winter, 23 April for spring, 26 July for summer and 6 October for autumn.

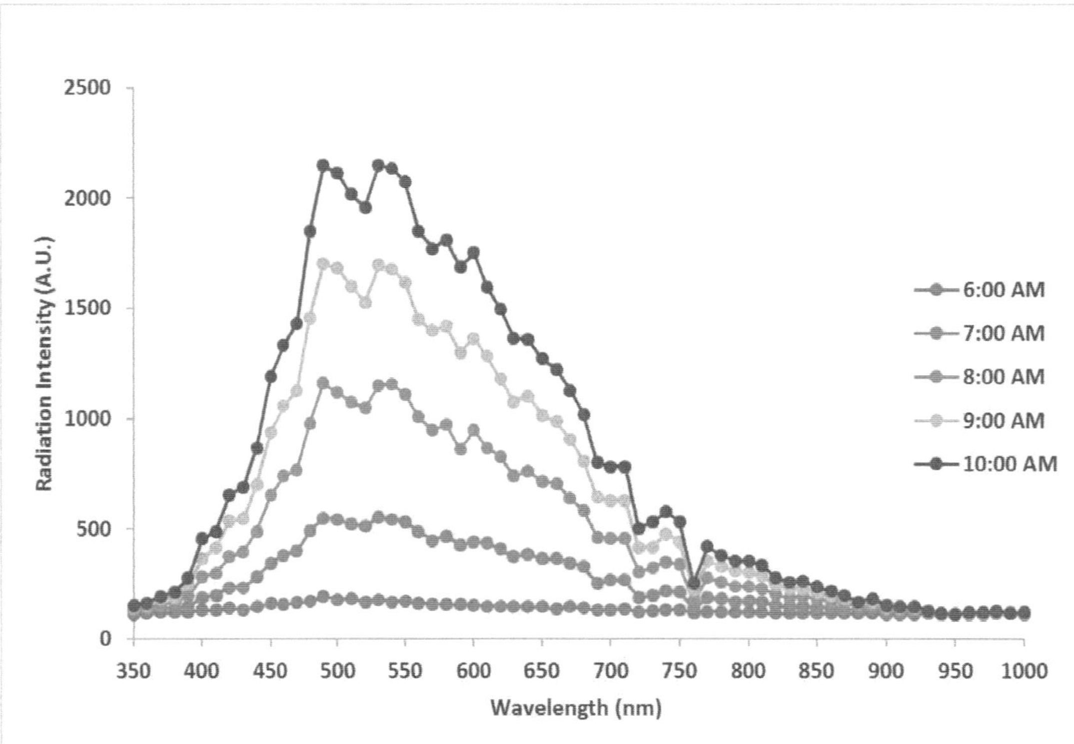

Figure 1 Measured solar spectral irradiance variation versus time (A.Ms) at Helwan on 17 September 2011.

For each of the 12 days selected for the analysis, the daily variation of ln I_λ at the six wavelengths is presented. On all 12 days, ln I_λ is analyzed at 4-5 air mass optical thickness. For the investigation of ln I_λ trend before and afternoon, Figure.2 and 3 have been plotted. Figure 2 shows, an example of the resultant plot data obtained befornoon time as a sampling of respective fitted lines for represented spectral bands in the UV (350,368 nm), visible (500, 615 and 780 nm) and near infrared (870 nm) regions. As expected for 21 January, the pattern of ln I_λ follows a decreasing trend before noon at all wavelengths strongly depending on SZA, where an exponential decrease curve. This behavior depicts that with increasing the air mass, the attenuation of the spectral solar radiation at the six selected wavelengths received by the earth also increases which closely related to the spectral changes due to the path through the atmosphere that cause attenuation (scattering and absorption of sun light). i.e. the attenuation coefficient α_λ increases. This trend was observed for all days of each month of the year, forming a characteristic pattern that occurs for all days. It is clear that wavelength 350 nm has the lowest decrease rate which means that the rate of change of its component function is nearly fixed all day long, while the decrease rate at wavelength 500 nm is the highest which indicates that the rate of change of its component function is very changeable all day long.

Above 70% of the correlation coefficient perceived was higher than 97% displayed by the least squares fit line. The correlation coefficients to fitted lines are $R^2 \geq 97\%$ on most days of measurements, which is very good correlation between the fitted lines and the collected data. These results confirm that there is stability in the local atmosphere during each half of the sampled days. As shown in Figure 3, it's found that at air mass 1, there is a difference between before and afternoon values of ln I_λ whereas the air mass increases, the difference between them increases.

For extinction values of the four seasons represented by the four clear selected days in table 1, there are significant differences between before and afternoon values at the six wavelengths where the afternoon values are higher than that of the before noon. These differences caused by the wind direction which transports all pollutants from Turah acment and iron steel factories from the east northern to the west side where the sunset. As a result, it has been found that the east side is clear relative to the west side, so there is a high opacity and low visibility in the west direction. Moreover, there are significant variations of extinction values in each season can be also observed from the data in table 1.

IOP Conf. Series: Materials Science and Engineering 1269 (2022) 012011 doi:10.1088/1757-899X/1269/1/012011

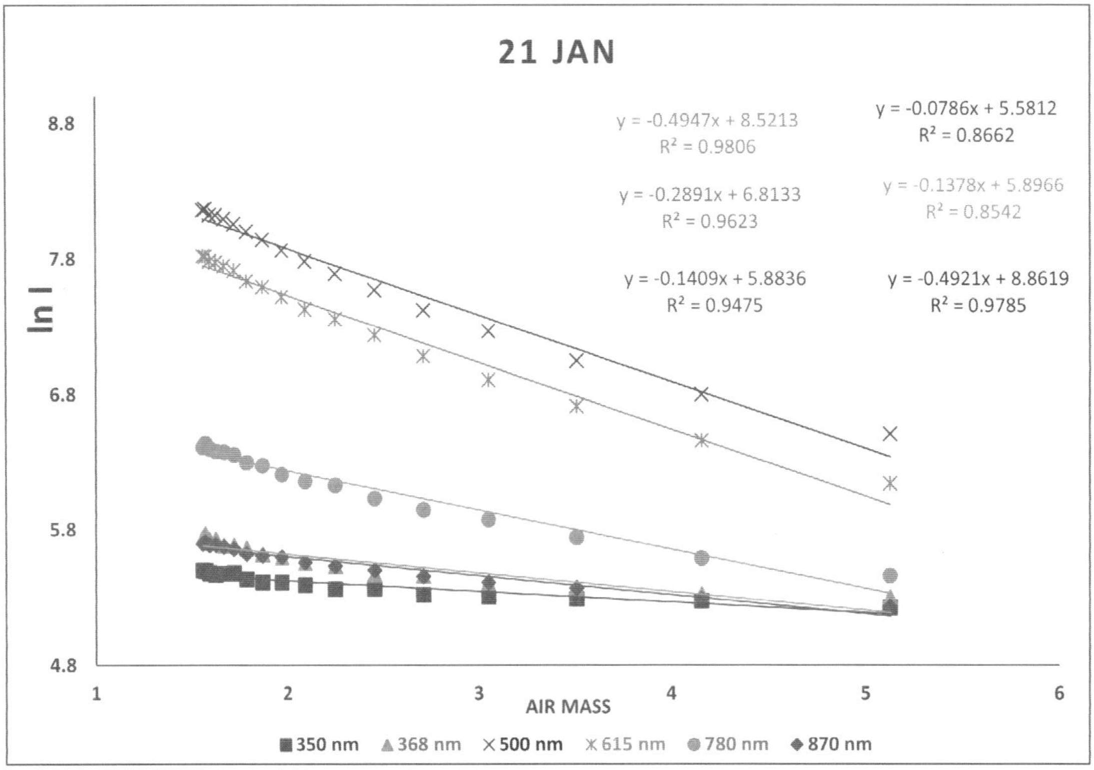

Figure 2 Langley plot for the intensity of the six selected wavelengths and air mass on 21 January 2011 before noon.

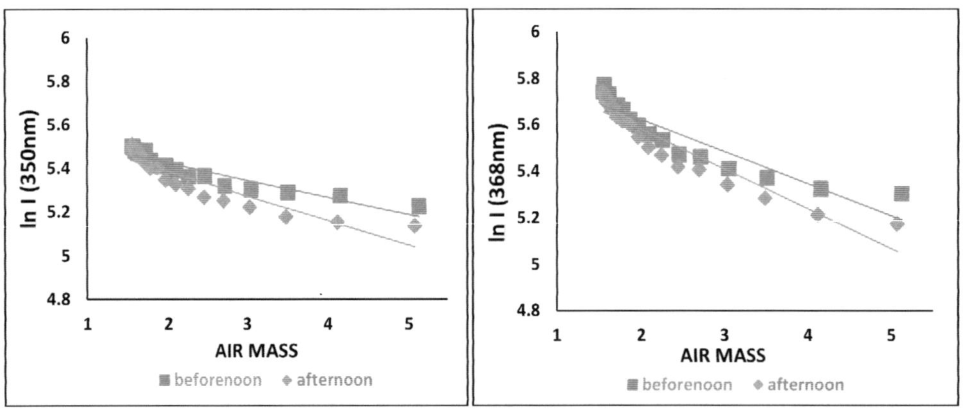

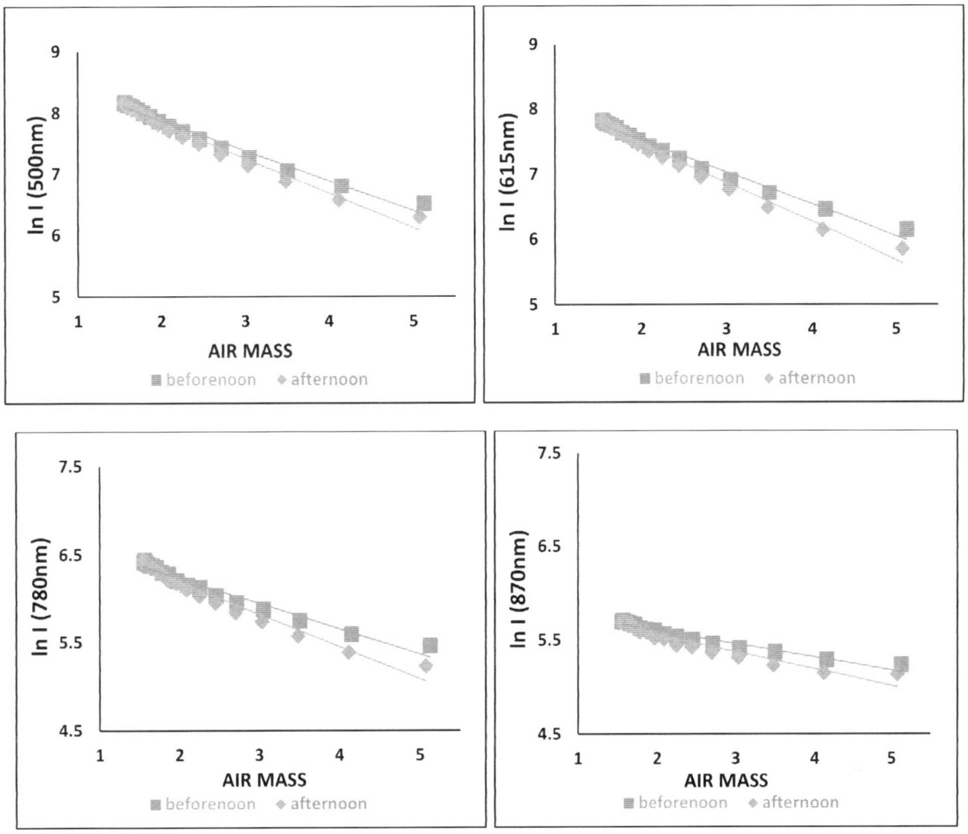

Figure 3. Variation of ln I_λ at the selected wavelengths with air mass for 21 January 2011 before and after noon.

Table 1. Extinction difference between before and afternoon values of the four seasons represented by the clear days (21 Jan, 23 April, 26 July, 6 Oct).

Wavelength (nm)	Winter (Δ α)	Spring (Δ α)	Summer (Δ α)	Autumn (Δ α)
350	0.03	0.03	0.005	0.03
368	0.03	0.02	0.02	0.03
500	0.08	0.2	0.19	0.14
615	0.1	0.19	0.2	0.16
780	0.08	0.12	0.13	0.11
870	0.04	0.07	0.08	0.05

IOP Conf. Series: Materials Science and Engineering 1269 (2022) 012011 doi:10.1088/1757-899X/1269/1/012011

Beginning from 21 January, seasonal variation of ln I_λ is presented in Figure 4 at the six wavelengths. Figure 4 shows the variation of ln I_λ of the four seasons at the six selected wavelengths with air mass before noon. The preliminary results obtained indicate high variability in both UV and IR spectral regions of wavelengths (350, 368 nm) and 870 nm, respectively, between extinction coefficients of the selected days that represent each season. On the other hand, for the visible spectral region of wavelengths (500, 615 and 780 nm) there are minor differences between extinction values of the selected days. As it is clear from Figure 4 the two wavelengths 500 nm and 615 nm have the same difference in α through the four seasons. This is indicating that, their components are function in them and not change in their concentration along the four seasons. Otherwise the wavelengths 350 nm, 368 nm and 870_nm have great differences. This may be due to the permanent effects of ozone absorption in the extinction of the visible spectrum especially from 450 to 770 nm. Alternatively, in case of UV and near IR spectral regions of wavelengths 350, 368 nm and 870 nm,

The ln I_λ curve has a variation through a day with particular turbidity levels different from that for another day with altered atmospheric conditions. In general, the changes in atmospheric conditions, aerosol load and properties present over the year particularly before and afternoon can be detected through the ln I_λ plot against air mass for four clear selected days at the selective wavelengths.

.

IOP Conf. Series: Materials Science and Engineering 1269 (2022) 012011 doi:10.1088/1757-899X/1269/1/012011

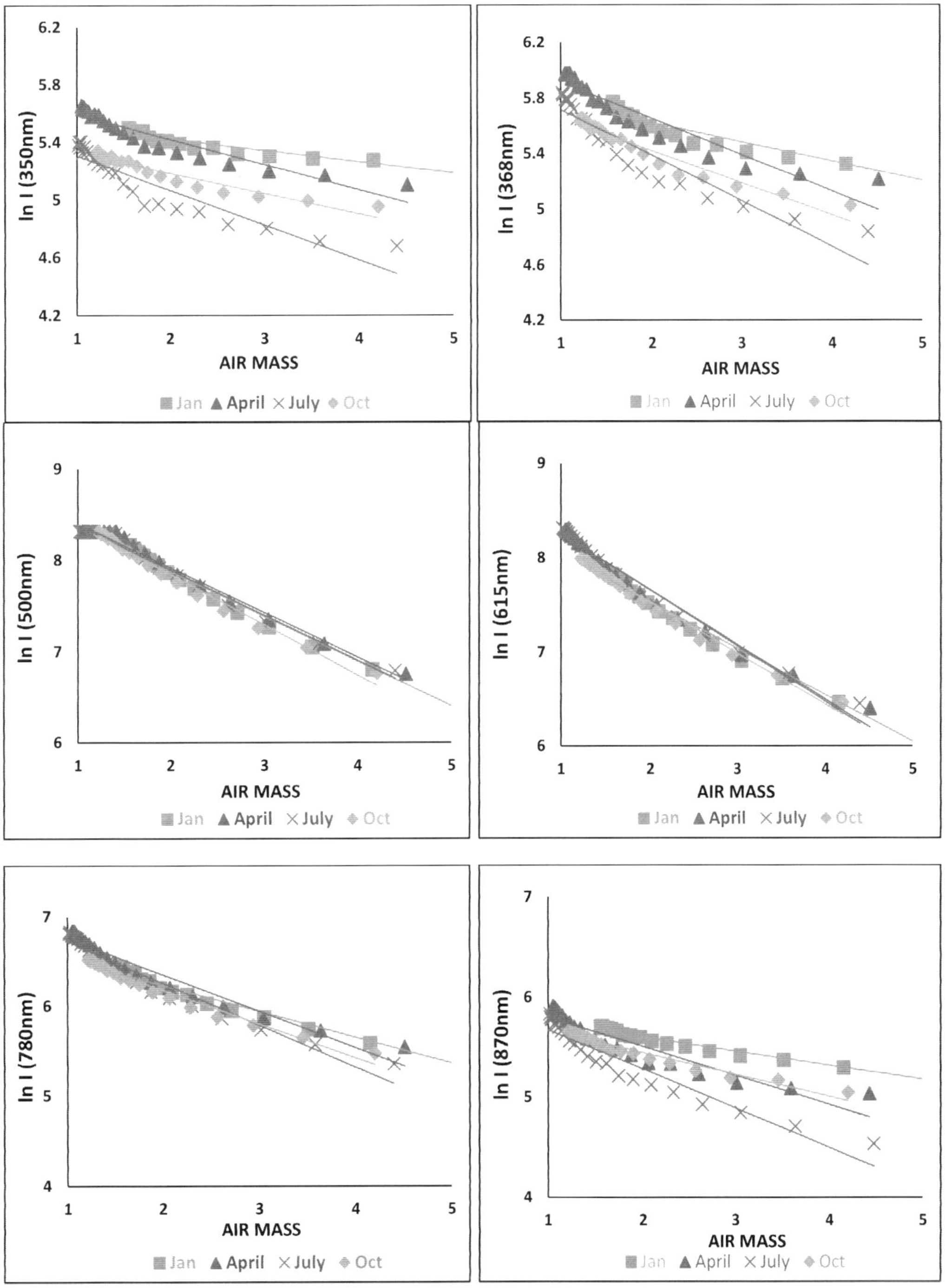

Figure 4 Seasonal variation of ln I_λ at six selected wavelengths with air mass before noon.

4.2. Calculation of the extinction coefficient (α)

During the period of study, all collected data has been fitted to a linear function (least-square approach), independently for the morning and afternoon data. The negative slope of the fitted line result of the Langley plot has been used for obtaining the total atmospheric extinction α [13]. The total atmospheric extinction coefficient result of the Langley plot for the three spectral regions of UV, visible and near IR is displayed in table 2. It presents the daily values of the extinction coefficient and shows that there is a variation can be observed from the data of extinction values in each month. In case of before noon, extinction values for wavelength 350 vary between 0.14 km^{-1} and 0.16 km^{-1} from March to September and then decrease to 0.06 km^{-1} in December, corresponding to the rainy season. Rain can make the sky clearer from dust [14].

Inspecting the results in table 2 it is apparent that, spring values of (α) are close to summer values, whereas the pronounced difference between summer and winter values may be attributed to the seasonal variation in water vapor in the atmosphere. Comparable patterns of variation to those of wavelength 350 nm can be obtained for the other wavelengths. The extinction coefficients for clear days generally increases from spring to summer with a maximum peak in July (0.24, 0.33 km^{-1}) for wavelengths 350 and 368 nm respectively, also the wavelengths 615, 780 and 870 nm have peak values in July (0.59, 0.47, 0.31 km^{-1}), respectively. In contrast to these wavelengths, it was found that wavelength 500 nm has a peak value of extinction coefficient in October.

Furthermore, it is clear from table 2 that the extinction values remain high during the autumn and spring seasons showing a peak in September and April, for wavelengths 350, 368, 500, 615,780 and 870 nm, respectively. The large extinction values in the transition seasons are due to the peaks in Particulate matter less than 10 μm (PM_{10}) especially Si silicon, and calcium, Ca, which are regarded as the main components of PM_{10} measured in the considered site. Moreover, elements related to fly ash, as well as the peaks in surface O_3 and NO_2. Desert dust is extra significant air pollution source. In spring, strong winds drive a lot of desert particles and transport them over most areas of Helwan towards the west. The low values in winter may be attributed to the replacement of the summer boundary layer laden through dust over Helwan via the cold air masses upcoming from the Mediterranean region. From table 2 we noticed that the maximum values of extinction coefficient at the wavelengths (350, 368, 500, 615, 780 and 870 nm) recorded in the summer season and minimum in winter due to high concentration of suspended particles in summer and minimum in winter, the wavelength 350 nm has the lowest value of extinction coefficient since attenuation is very strong in the far-ultraviolet region and it diminishes sharply with wavelength until, at λ~350 nm, ozone becomes almost transparent to solar radiation. The extinction coefficient increases again in the visible region and reaches a maximum value at λ=615 nm, therefore the wavelength 615 nm has the highest value of extinction coefficient because ozone has a strong absorption band in the visible region from 450 to 770 nm. More else the high levels of surface ozone O_3 that absorbs much in the summer months, and became highest when transported by wind from an iron and steel factory in the considered site.

IOP Conf. Series: Materials Science and Engineering 1269 (2022) 012011 doi:10.1088/1757-899X/1269/1/012011

Table 2. The variation of the extinction coefficient (α) for 12 clear days, beforenoon (b) and afternoon (a) at the six selected wavelengths.

Clear days	αb (350)	αb (368)	αb (500)	αb (615)	αb (780)	αb (870)
21-Jan	0.07	0.1378	0.4921	0.494	0.289	0.1409
18-Feb	0.083	0.1275	0.458	0.456	0.275	0.1338
17-Mar	0.142	0.2157	0.5251	0.5829	0.3831	0.207
23-Apr	0.175	0.2611	0.4886	0.582	0.405	0.2233
16-May	0.171	0.2579	0.4962	0.562	0.39	0.2257
22-Jun	0.1997	0.2803	0.5054	0.58	0.43	0.2691
26-Jul	0.2407	0.3282	0.4941	0.5905	0.4674	0.3071
18-Aug	0.1808	0.2639	0.5042	0.578	0.4204	0.258
17-Sep	0.1569	0.2415	0.5191	0.5578	0.3925	0.2268
6-Oct	0.1411	0.2306	0.5531	0.5434	0.3753	0.2188
8-Nov	0.1045	0.1697	0.5282	0.523	0.3292	0.1576
11-Dec	0.0696	0.1131	0.4861	0.4796	0.2564	0.112
Clear days	**αa (350)**	**αa (368)**	**αa (500)**	**αa (615)**	**αa (780)**	**αa (870)**
21-Jan	0.112	0.1704	0.5679	0.596	0.365	0.1801
18-Feb	0.124	0.1792	0.5794	0.583	0.353	0.1872
17-Mar	0.175	0.2537	0.6388	0.7004	0.4581	0.2548
23-Apr	0.2	0.2807	0.6847	0.774	0.529	0.2908
16-May	0.192	0.2856	0.7023	0.761	0.516	0.2965
22-Jun	0.211	0.2862	0.6479	0.729	0.531	0.3315
26-Jul	0.2451	0.3474	0.6803	0.7956	0.5998	0.3913
18-Aug	0.222	0.3223	0.7045	0.7931	0.588	0.3599
17-Sep	0.1755	0.2524	0.6275	0.6882	0.4863	0.2858

6-Oct	0.1698	0.2565	0.6884	0.6995	0.4815	0.2725
8-Nov	0.1404	0.2094	0.6845	0.721	0.4575	0.2442
11-Dec	0.095	0.1375	0.5582	0.5746	0.3271	0.1655

Figure 5 shows the relation between extinction coefficient (α) and wavelength (λ) for before and afternoon. The extinction value starts to increase in the visible spectral region from 450 nm and reaches the maximum at 615 nm, then begins to decrease with increasing wavelength, i.e., the extinction coefficient is inversely proportional to the radiation wavelength. It means that when the radiation wavelength decreases the absorption and scattering increases, owing to aerosols existing in the atmosphere. These experimental results (for the visible range: 615–870 nm) corroborate the findings reported by Cerqueira et al. (2014) [4] for the bands 500, 670 and 870 nm, and Aljawi et al. (2016) [6] in the range 400-700 nm. The increase in the atmospheric extinction of the visible spectrum at (500 and 615 nm) may be also attributed to the Rayleigh scattering in which the particle size is a function in the wavelength λ as: $\pi d/\lambda$, where d is the particle size. Rayleigh scattering is fulfilled when $\pi d/\lambda < 0.6/\pi$ [9]. In more details, it has been found that the particle size at Helwan site ranged from 200-300A [15]. This range of particle size scatter practically any visible wavelength namely wavelengths of 500, 615 nm effectively in the considered site and fulfill the criteria of Rayleigh scattering [16].

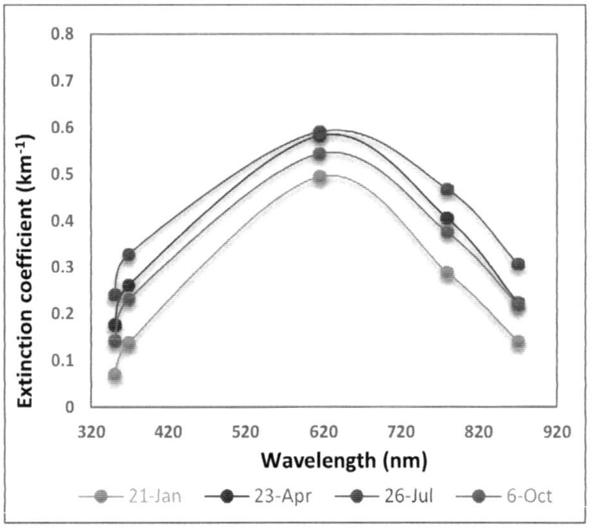

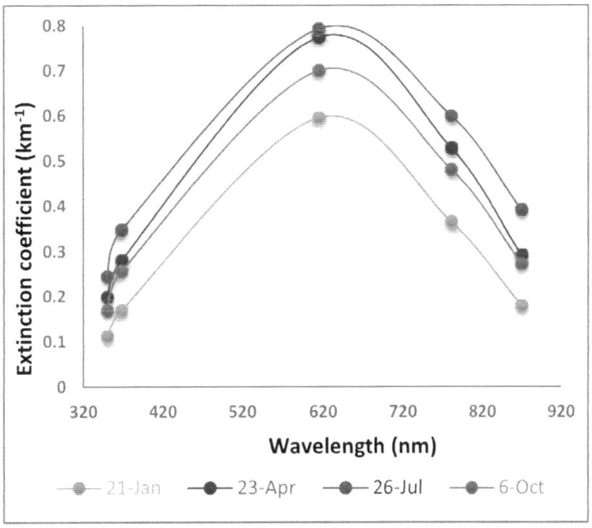

Figure 5 Extinction coefficient variation for 12 clear days with wavelength before noon (up) and afternoon (down), respectively.

4.3. Calculation of the atmospheric transparency (T)

Atmospheric turbidity which affects the atmospheric transparency depends on the local weather conditions and the climate of the site. The prevailing winds may transport aerosol particles from distant sources, play a major role in the temporal variation of turbidity and consequently the transparency. This behavior was reported by Hassan (1995). He found that the N and N-E is the main source of wind direction (50% from total direction), in this sense, both Helwan Portland and Tura cement factories signify 50% of the pollution of Helwan region. The national cement factory and Iron and steel factories represent about 40% (N-W, S, W, S-W directions) of the total of pollution of the Helwan region. The levels of O_3 and NO_2 were highest when transported from an iron and steel factory is located. Also El-Tebeen (S-W from Helwan) is considered a very high polluted region because all cement factories 50% from the Iron and steel factory in this region. Hence the extinction coefficient for the afternoon is greater than that of before-noon. This leads to higher values of transparency before-noon and lower values in the afternoon [17].

Additionally, the major days in the year at Helwan, the east side is clear as the relative. Hence there is a high opacity and low visibility in the west direction. This is due to the increase in the afternoon temperature which leads to an expansion in the lower layers of the atmosphere hence an increase in the thickness of the layers so we have greater cross sectional area for the suspended particles in the atmosphere. Moreover, the high temperature makes an excitation for the suspended particles that present in the atmosphere.

Additional main factor that could affect the transparency of the atmosphere to the solar radiation is the wind direction. Other than the diurnal variation of the atmospheric transparency factor, represented by the difference between the morning and after- noon values, a strong seasonal variation is seen in table 3, with a maximum in winter and a minimum in summer for the wavelengths (350,368, 615,780 and 870 nm). It was found that, the highest values of transparency were recorded in winter resulted from the increase in relative humidity and the decrease in temperature. In this sense, the water uptake is increased in the aerosols and consequently both particle size and mass increase as well as the deposition velocities. . In summer, the main sourece of aerosols is the photochemical processes this directly leads to reduced valuesof atmospheric transparency. This is in a good agreement with the previously published work by Rahoma (1998) [18]. Rahoma et al demonstrates that high concentration of suspended particles in summer (1200-1800 μg/m3) and minimum in winter (600-840 μg/m3) with respect to the industrial area. The maximum in summer (June, July and August) at the high temperature, minimum relative humidity, and faster wind speed are assisted to take place dispersion in the suspended particles. Hence, they cause large heating in the lower layer in atmospheric contents. For both parameters affecting heating in the lower layer which contains the majority of dust particles are caused that; the back scattering alone induces heating above the layers of dust, as reflected radiation increase to the higher layer of dust. Table 3 shows the transparency values of the selected 12 clear days. It is apparent that the maximum values of transparency at the wavelengths (350,368, 615,780 and 870 nm) recorded in the winter season and minimum in summer. The wavelength 615 nm has the lowest value of transparency, whereas the wavelength 350 nm has the highest value of transparency. As it is clear in Figure 6 Transparency values in the afternoon are lower than before noon ones and they are shifted toward the longer wavelengths. This may be explained by the fact that as the solar elevation decreases the intervening air mass increases and the minimum of the solar spectrum is shifted toward the longer wavelengths.

Table 3. The variation of the atmospheric transparency (T) for 12 clear days before and afternoon at the six selected wavelengths.

Clear days	Tb (350)	Tb (368)	Tb (500)	Tb (615)	Tb (780)	Tb (870)
21-Jan	0.932	0.871	0.611	0.610	0.749	0.869
18-Feb	0.920	0.880	0.633	0.634	0.760	0.875
17-Mar	0.868	0.806	0.591	0.558	0.682	0.813
23-Apr	0.839	0.770	0.613	0.559	0.667	0.800
16-May	0.843	0.773	0.609	0.570	0.677	0.798
22-Jun	0.819	0.756	0.603	0.560	0.651	0.764
26-Jul	0.786	0.720	0.610	0.554	0.627	0.736
18-Aug	0.835	0.768	0.604	0.561	0.657	0.773
17-Sep	0.855	0.785	0.595	0.572	0.675	0.797
6-Oct	0.868	0.794	0.575	0.581	0.687	0.803
8-Nov	0.901	0.844	0.590	0.593	0.719	0.854
11-Dec	0.933	0.893	0.615	0.619	0.774	0.894
Clear days	**Ta (350)**	**Ta (368)**	**Ta (500)**	**Ta (615)**	**Ta (780)**	**Ta (870)**
21-Jan	0.894	0.843	0.567	0.551	0.694	0.835
18-Feb	0.883	0.836	0.560	0.558	0.703	0.829

IOP Conf. Series: Materials Science and Engineering 1269 (2022) 012011 doi:10.1088/1757-899X/1269/1/012011

17-Mar	0.839	0.776	0.528	0.496	0.632	0.775
23-Apr	0.819	0.755	0.504	0.461	0.589	0.748
16-May	0.825	0.752	0.494	0.467	0.597	0.743
22-Jun	0.810	0.751	0.523	0.482	0.588	0.718
26-Jul	0.783	0.707	0.506	0.452	0.549	0.676
18-Aug	0.801	0.724	0.495	0.451	0.555	0.698
17-Sep	0.839	0.777	0.534	0.502	0.615	0.751
6-Oct	0.844	0.774	0.502	0.497	0.618	0.761
8-Nov	0.869	0.811	0.504	0.486	0.633	0.783
11-Dec	0.909	0.872	0.572	0.563	0.721	0.847

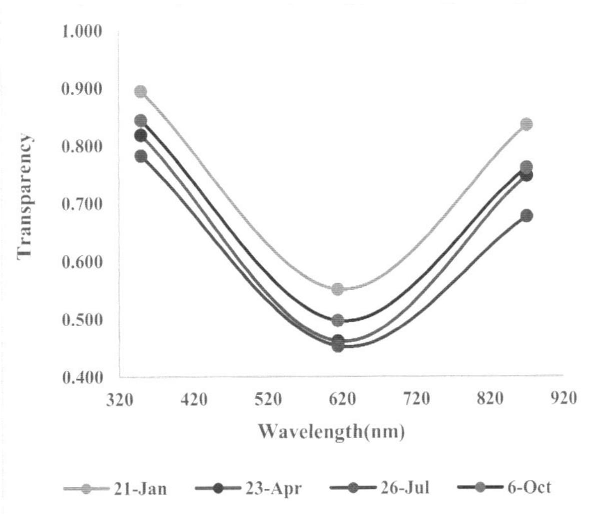

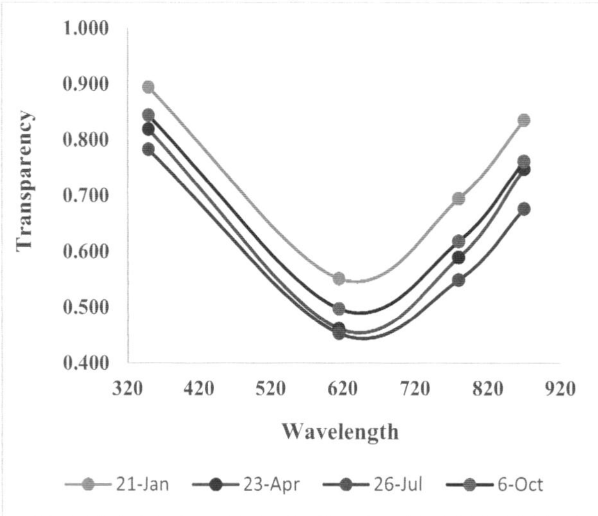

Figure 6 Transparency variation for four seasons with wavelength before noon (up) and afternoon (down), respectively.

5. Conclusion

Twelve clear-sky days with Continuous, measurements of solar spectral irradiation using the Spectroradiometer have been achieved to analyze the diurnal variation of extinction coefficient and atmospheric transparency at Helwan, in Egypt. The extinction coefficient and atmospheric transparency are shown to be the most significant parameters affected by the air mass. It is quite characteristic that the diurnal variation of extinction coefficient exhibited a similar pattern corresponded to all wavelengths with higher values in the afternoon. The extinction coefficient depicts obvious diurnal variations in each season proportionally to the wavelengths. High regression factor is obtained in the summer that means the extinction is more expected in this season owing to aerosol concentrations. Summer and winter show the highest and lowest extinction coefficients, respectively for all wavelengths under investigation. It has been observed that the highest value of extinction coefficient occurred at 615 nm, while the lowest one taken place at 350 nm. The line fitted to the data show very high coefficients of correlation, with maxima of 99.26%, 99.1%, 98.18 % and 95.53% in the bands 500, 615,780 and 870 nm, respectively during the period of study. In contrast, it was found that transparency values in the afternoon are lower than before noon ones and the highest values predictable in winter whereas the lowest values occurred in the summertime. In addition, it has been obtained that transparency values were shifted toward the longer wavelengths in the case of afternoon data. Finally, The atmospheric transparency (T) of the six wavelengths under consideration satisfy the following inequality T (350)> T(870) > T(368) > T(780) >T(500) > T(615).

References

[1] Yang L., Gao, X., Li, Z., Jia, D. (2022). *Energy Reports*, *8*, 1935-1948.

[2] Giavis, G. M., Kambezidis, H. D., Sifakis, N., Toth, Z., Adamopoulos, A. D., Zevgolis, D. (2005). *Atmospheric research*, *78*, 79-92.

[3] Gutiérrez-Trashorras A. J., Villicaña-Ortiz E., Álvarez-Álvarez E., González-Caballín J. M., Xiberta-Bernat J., Suarez-López M. J. (2018). *Renewable and Sustainable Energy Reviews*, *81*, 93-106.

[4] Kocifaj, M., (2016). Journal of Quantitative Spectroscopy and Radiative Transfer, *173*, 20-25.

[5] Cerqueira, J. J. G., Fernandez, J. H., Hoelzemann, J. J., Leme, N. M. P., Sousa, C. T. (2014). *Advances in Space research*, *54*, 1530-1543.

[6] Che, H., Zhang, X., Li, Y., Zhou, Z., Chen, Z. (2006). *China Particuology*, *4*, 327-329.

[7] Aljawi, O., Gopir, G., Kamil, W. W. M., Mohamad, N. S. (2016). *AIP Conference Proceedings*, *1784*, 040004.

[8] Hanrieder, N., Wilbert, S., Schroedter-Homscheidt, M., Schnell, F., Guevara, D. M., Buck, R., Giuliano, S., Pitz-Paal, R. (2017). *AIP Conference Proceedings*, *1850*, 140011.

[9] Khalil, S. A., Hassan, A. H., Rahoma, U. A., Abulwfa, A., Khamees, A. S. (2021). *Open Journal of Applied Sciences*, *11*, 294-311.

[10] Liou, K.N. (2002). *An introduction to atmospheric radiation*, *84*, 1-583.

[11] Reagan, J.A., Thome, K., Herman, B., Gall, R. (1987). In: Proc. *International Geoscience and Remote Sensing* Symposium, 63e67.

[12] Bruegge, C. J., Conel, J. E., Green, R. O., Margolis, J. S., Holm, R. G., Toon, G. (1992). *Journal of Geophysical Research: Atmospheres*, *30*, 18759-68.

[13] Ichoku, Wei, P. S., Hsieh, Y. C., Chiu, H. H., Yen, D. L., Lee, C., Tsai, Y. C., Ting, T. C. (2018). *Heliyon*, *1*, e00785.

[14] Carra, E., Ballestrín, J., Monterreal, R., Enrique, R., PoloJ., Fern´andez-Reche, J., Barbero, J.,Marzo, A., Alonso-Montesinos, J., L´opez, G., Díaz, B. (2022). *Sustainable Energy Technologies and* Assessments, *51*, 101991.

[15] Rhoma , U.A., Khalaf, G. M., G., Hashem, H. M., El-Hagary, M. (2016). *International Sixth Conference Sustainable development and Arab Economic security, At Institute of Environmental Studies and Research Ain Shams University, Abbasia, Cairo, Egypt*, 34.

[16] Bernardo, V., Pinto, L. J., Schade, U., Rodriguez-Perez, M. A. (2020). *Colloids and Surfaces A: Physicochemical and Engineering Aspects*, *600*, 124937.

[17] Hassan, A.H., Ph.D, (1995).

[18] Mosalam Shaltout, M. M., Hassan, A. H., Rahoma, U., A. (1998). *In Proceeding of 6th international* conference on energy and environment, Cairo, Egypt, *2*, 567-578.

AUTHOR INDEX

Institute of Physics
Dirac House, Temple Back
Bristol BS1 6BE UK

ISSN: 1757-8981
ISBN 978-1-7138-6428-8

9 781713 864288

CURRAN ASSOCIATES INC.
proceedings.com